BUILDING
HIGH PERFORMANCE
BUSINESS
RELATIONSHIPS

*Rescue, improve and transform
your most valuable assets*

TONY LENDRUM

WILEY

John Wiley & Sons Australia, Ltd

First published 2011 by
John Wiley & Sons Australia, Ltd
42 McDougall St, Milton Qld 4064

Office also in Melbourne

Typeset in ITC Berkeley Oldstyle Std 10.75/14.5pt

National Library of Australia Cataloguing-in-Publication data:

Author:	Lendrum, Tony.
Title:	Building high performance business relationships: rescue, improve and transform your most valuable assets / Tony Lendrum.
ISBN:	9780730377740 (pbk.)
Notes:	Includes index.
Subjects:	Business communication.
	Success in business.
	Strategic planning.
Dewey Number:	650.13

Cover design by Peter Reardon, Pipeline Design <www.pipelinedesign.com.au>

Cover image: ©iStockphoto.com/Stephen Strathdee

Printed in China by Printplus Limited.

10 9 8 7 6 5 4 3 2 1

Disclaimer

The material in this publication is of the nature of general comment only, and neither purports nor intends to be advice. Readers should not act on the basis of any matter in this publication without considering (and if appropriate, taking) professional advice with due regard to their own particular circumstances. The author and publisher expressly disclaim all and any liability to any person, whether a purchaser of this publication or not, in respect of anything and of the consequences of anything done or omitted to be done by any such person in reliance, whether whole or partial, upon the whole or any part of the contents of this publication.

This book is dedicated to Julie, my wife and business partner for more than 30 years, and my two sons, Simon and Mark, for their continuous love and support.

CONTENTS

ABOUT THE AUTHOR

Tony Lendrum is founder and director of Strategic Partnering Pty Ltd, a management consulting firm he formed in September 1994 to work with organisations interested in pursuing the benefits of strategic partnering and alliance relationships, and high performance relationship management. Tony has established himself globally as a recognised authority, facilitator and coach in these fields.

Tony Lendrum is the author of two highly successful books, *The Strategic Partnering Handbook,* now in its fourth edition, and *The Strategic Partnering Pocketbook*. His books are widely used by small, medium and large organisations in both the private and public sectors to provide a model and guide for implementing effective partnering and alliance strategies. His books are benchmarks for relationship management generally.

Tony is also the founder and director of 0 to 10 Relationship Management Pty Ltd, which offers training programs in the practical application of 0 to 10 Relationship Management™ (0 to 10RM), online business relationship health check diagnostics, and the licensing of 0 to 10RM elite trainer facilitators (ETFs). ETFs form communities of relationship champions, working independently and together to create value for their own and other organisations.

Tony's work spans more than 30 years, 14 of which were spent with ICI, the global chemical company. Tony holds an honours degree in physical chemistry, and he has worked in many technical, sales, marketing, management, business development and manufacturing roles. He spent two years in South Korea as ICI's business development manager with responsibility for acquisitions, joint ventures and new business opportunities. He was also operations manager for one of ICI Australia's petrochemical plants.

ACKNOWLEDGEMENTS

My thanks go to those stubbornly passionate innovators and champions with whom I have worked over the past 20 years—those people who have an uncompromising belief in the value and importance of relationships. The list is endless, but in particular I would like to thank Jock Macneish for his boundless energy, powerful images and challenging intellect. Jock's creative images, spread throughout the book, give insight and bring stories to life. Sometimes the messages will be as clear as day, sometimes hidden like treasure. I have conducted workshops in Hong Kong and mainland China in Chinese, but I don't speak Chinese! The pictures did their own facilitation, spoke their own words and told their own stories. They bridge the gaps in culture, language and location across all generations.

INTRODUCTION

Success in business is critically linked to the quality and performance of both internal and external relationships. If people are an organisation's greatest asset, then the relationships they form are a prime indicator of their quality and performance, and therefore a lead indicator of business success and sustainability. Evolving business relationship management as a core competency will give an organisation a key strategic platform upon which superior performance, cultural transformation and competitive advantage can be sustained.

While the focus of this book is on business relationships, there is a direct relevance to relationships in our personal lives and throughout society as a whole. Mastering the art of relationship management will have a direct and positive impact on personal insight, individual empowerment, work–life balance and quality of life.

0 to 10 Relationship Management (0 to 10RM) comprises a methodology and approach for managing the relationship journey of improvement from the current state to the desired future state. The 0 to 10RM methodology and approach applies to the relationships with a single customer or supplier or stakeholder through to the development of a high-level relationship strategy involving many relationships, with varying current states and differing desired future states. Relationships in this context can be defined as those human associations, connections or interactions — real or virtual — that have a goal or a purpose. The 0 to 10 Relationship Management Matrix (see chapter 1) provides the framework for understanding and implementing relationship rescue, improvement and transformational strategies.

The 0 to 10RM relationship scale was first presented in the first edition of *The Strategic Partnering Handbook*. The focus at that time was on partnering and alliance relationships. However, business relationships come in all shapes and sizes and, as it turns out, not all relationships need to be partnerships or alliances. It is the full spectrum of 0 to 10 relationship types and associated performance levels that need to be understood and managed. An understanding of the spread of relationships (both type and performance levels) is required to support business goals and objectives.

A range of both current states and desired future states exist across internal and external customer or supplier relationships, and so there are a number of different relationship improvement journeys to manage.

Many of the current 0 to 10RM tools and models are unaltered from those described in my previous books, *The Strategic Partnering Handbook* and *The Strategic Partnering Pocketbook*. Others are variations and modifications of those models and tools. Experience has taught me that they all have practical application right across the 0 to 10RM spectrum, and don't apply just to the partner segment as first envisaged. This book is a natural evolution from my earlier books: adapting existing material, adding new material, and expanding the conversation from a core focus around partnerships and alliances to relationship management more broadly.

0 to 10RM is people-focused and principle-centred, and it applies across the full gamut of human relationships. The models, tools, applications, checklists and diagnostics in this book are as relevant to directors in the boardroom as they are to sales, marketing and procurement professionals at the coalface, shopfloor and operating levels, and all people in between. This book will help turn theory and rhetoric into successful practice. It provides a framework for common understanding, common language and common practice around business relationship management in an increasingly complex and fast-paced global environment.

A storyboard and story-telling style is used throughout this book to create narratives around which business relationships can be better understood and managed. The underpinning six principles and five themes together form the 0 to 10RM storyboard (see chapter 1), which presents a single integrated picture for 0 to 10RM. By reading and applying the material in this book you will be able take any conversation and place it somewhere on the storyboard, tell a story, generate discussion and understanding, challenge and innovate for improvement, and add value for your organisation. The 0 to 10RM storyboard, the six principles and five themes are introduced in chapter 1.

The book is divided into three parts — A, B and C — which reflect the three sections of the 0 to 10RM storyboard. At the end of each chapter a summary outlines the major topics and key points in the chapter. The reader can either read the book sequentially or quickly identify areas of interest from the summary and focus on the detail.

Part A (chapters 2 to 4) introduces the framework and components of 0 to 10RM. Chapter 2 looks at the 0 to 10RM Matrix and details the diagnostic tools used to understand and evaluate the current state and desired future state of relationships, and discusses which relationship types are appropriate in different circumstances. Chapter 3 discusses the 11 legitimate relationship types and performance levels.

Chapter 4 analyses the seven key components that need to be considered and managed, irrespective of the relationship type and performance levels reached.

Part B (chapter 5) discusses in depth the Let's Go change model, along with the role competencies for 0 to 10RM high performance relationship managers. It also looks at people, high performance culture and the effective management of change in terms of engaging and sustaining the right attitudes and behaviours required to deliver the desired results from business relationships.

Part C (chapters 6 and 7) considers managing the journey from the current state of relationships to the desired future state. Chapter 6 reviews the relationship development curve presenting the connection between time, the quality of relationship outcomes, and the different stages and phases engaged throughout the improvement journey. Chapter 6 also introduces the concept of relationship journey management and the journey to be taken in bridging the gap between the current state(s) and desired future state(s). It addresses a number of matters: how to sustain high performance relationships and deliver superior value beyond the life of key people; how to engage effective relationship rescue, relationship improvement and relationship transformation strategies; and the phases of relationship development and how long they will take. Paradigm shifting, innovation and continuous improvement are key areas of focus. Chapter 7 considers the process of startup, as well as the ongoing management and improvement of relationships. It looks at the 12 motivators that drive the improvement process, the 12 steps that can be taken to implement and execute the journey, and the six outcomes that need to be considered and delivered on.

I hope that this book will provide enjoyment, insight and value, and will turn relationship management from a business competency into a life competency. I prefer high performing cooperative and collaborative relationships, be they internal or external, or at an organisational, team or individual level. This book and my work over the past three decades in the area of strategic partnerships and alliances, pioneering and community relationships support that preference. The reality is, however, that relationships come in all shapes and sizes, with varying degrees of alignment, understanding and performance. They don't always go the way we plan. There are controllables and non-controllables, different people and personalities, market forces, and local and world events that shape our relationships and destinies. To be successful in business and in life we need to understand and be able to effectively manage all types of relationships. 0 to 10 RM is a toolbox of principles, models and tools to do just that.

Your ability to interpret and apply the 0 to 10RM Matrix and the other 0 to 10RM principles and themes that make up the storyboard will benefit and add value for you and your organisation, and others.

CHAPTER 1
0 TO 10 RELATIONSHIP MANAGEMENT — A STORYBOARD PERSPECTIVE

Six principles and five themes underpin 0 to 10RM and, when combined into a single picture, they form the 0 to 10RM storyboard. The 0 to 10RM storyboard (shown in figure 1.1, overleaf) provides the canvas upon which to take a journey; a framework for telling stories; a set of engagement tools to simplify complexity; and a way to proactively manage and improve business relationships. This chapter gives an overview of the 0 to 10RM storyboard, leading to a more detailed analysis and understanding in the subsequent chapters.

I remember some years ago as an operations manager being frustrated at the lack of progress in implementing improvements in safety, reliability and productivity in a manufacturing plant. These improvements were critical for the ongoing viability and competitive advantage of the business. For unknown reasons there was significant resistance from the operations and maintenance teams. Bewildered, I approached Ian, one of the lead maintenance fitters, whom I had known for years and respected. I asked Ian, 'How is it that the operations and maintenance teams don't understand the importance of safety, reliability and productivity improvement for the sustainability of the business?' His response caught me by surprise, 'We know what you want and understand the importance of these things to the business. We also know how to deliver the benefits you are looking for but we're not going to do it'. 'Why not?' I asked, even more bewildered. 'We simply don't trust management to do the right thing by us when the improvements are delivered', he said. Then added, 'Many of us fear our jobs will be in jeopardy if the business improvements are made'.

A classic win–lose scenario. There was no trust and it had to be built. It was 12 months before annualised salary replaced the need for overtime and breakthrough, collaborative work practices were implemented. That simple conversation with Ian was a catalyst for fundamental change in the relationship between employees and management. His fear concerning job security turned out to be unfounded. A win–lose turned into a win–win.

Figure 1.1: the 0 to 10RM storyboard simplified

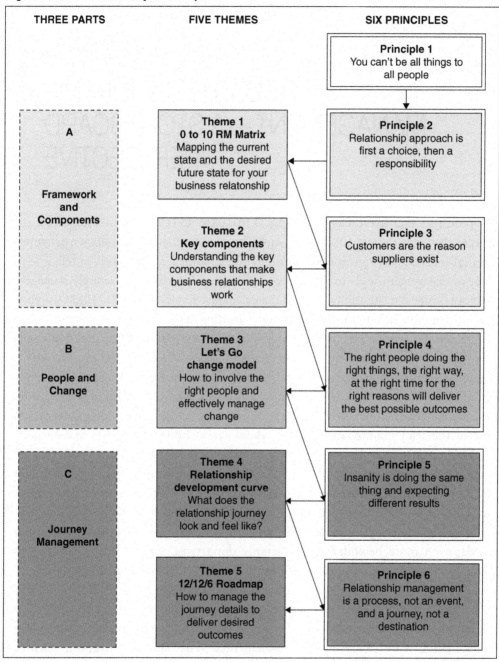

On another occasion, as a key account manager, I was attempting to convince a large strategic customer of the mutual benefits that could be achieved if they used our telemetry-based inventory management system, which was similar in principle to the technology used in modern business process outsourcing. It was slick technology in those days that would allow for the remote reading of customer silo levels; the numbers would then be transferred electronically to our manufacturing facility and automatically linked into our production schedules. We, as the supplier, would then control deliveries to the customer and drive efficiencies into our production process and their logistics and procurement systems. Simple, cost-effective and value-adding. A no-brainer in sales parlance. At least that is what I thought. The customer rejected the proposal outright, not only the first time but the second and third times. I raised the matter. The reason for the thumbs down was not the technology. It was that the customer didn't trust us with the management of a critical raw material and the associated, competitively sensitive, forecast data. They also wanted to continue to protect this information from us to trade us off against other suppliers for lower prices. Our relationship was simply not at a level that allowed the real benefits of such an arrangement to be mutually exploited. It took a year to implement the telemetry system with all its associated benefits. It took another two years for a true strategic partnership to be established around trust, transparency, shared risk, common goals and mutual benefit.

The underlying ideas in both stories are that people and trust are the basis of all relationships. Dig a little deeper and you will find an underpinning set of basic principles and a common suite of models and tools that can be used to manage the relationship improvement journey. These principles and tools are explored in detail in the following chapters.

Relationships are those human associations, connections or interactions that have a goal or purpose, and few people would argue the importance of effective, productive, successful and enjoyable relationships in business and in life generally. After all, life, liberty and the pursuit of happiness are founded on good relationships with others. So why are some relationships fragile and forever in crisis, while other relationships seem to go from strength to strength and are lasting, robust and mutually rewarding?

Relationship management is important because relationships — good, bad or indifferent — are central to the way business is managed. Better relationships result in better business and a better quality of life; this in turn makes our world a better place. Specifically, better business relationships will positively impact the six key result areas of (1) financial success; (2) customer/stakeholder satisfaction;

(3) sustainable competitive advantage; (4) best practice; (5) innovation; and (6) attitude.

While key people in the organisation are clearly critical assets, the irony is that high performance relationships must be enduring and successful beyond the individuals. Succession planning, the creation of new roles, the introduction of new management, the reality of employee turnover, the promotion of high achievers, mergers and acquisitions, and dealing with ever-changing market conditions will ensure there is a continual flow of people in and out of business relationships. Cultures, strategies, structures, processes and people that positively support the development and delivery of relationship value must be enduring and adaptive well beyond the first handshake.

The desired future state of relationships is the aspirational end point or major milestone we seek. It represents success through goals achieved along a journey undertaken. Different relationships may involve different products and services, different people and personalities, around different strategies and objectives. Each of these variables is dependent on or impacted by timing, location and environment. It is entirely appropriate that we aspire to and plan for those desired future states that are better than or provide more compelling alternatives to the current states. The need or desire for continuous improvement is integral to the human condition. Simply put, you can't achieve your goals unless you know where you are going and you have a plan to get there. It is, therefore, critical to begin the relationship improvement journey with the end in mind.

Even a status quo desired future state requires some degree of maintenance, flexibility and improvement to sustain a competitive position against the inevitable tides of market forces.

There can be no greater joy than completing a journey that turns a dream into reality. Envisioning desired future states requires the setting of goals. Achieving these goals provides us with success and a sense of achievement, a glimpse of the next horizon to be reached, and the will and commitment to take up the challenge with confidence and passion. Storytelling will bring these journeys to life and allow learning and knowledge to be passed on.

0 to 10RM—a storyboard perspective

The 0 to 10RM storyboard is as simple as A, B, C. That is:

A—Framework and key components of the relationship
B—People and change
C—Journey management

The 0 to 10RM storyboard presents a universal and value-adding set of principles, models and practical tools built up over almost 30 years of front-line experience. The 0 to 10RM principles, models and tools can be applied to all business relationships—large, medium or small—at all levels, and in both the public and private sectors. With the globalisation of markets now a reality, the world has never seemed so small, the marketplace so competitive, change more rapid and the pressure to perform so intense. Relationships have never been more important.

The 0 to 10RM storyboard (figure 1.1 on page 2) comprises three parts (A, B and C) presenting the six principles and five themes that make up the complete 0 to 10RM methodology and approach. The storyboard represents 0 to 10RM on a single page. All the details, applications and tools evolve from this single diagram. If you have any questions, want to engage a story or a conversation about the past, present or future, or implement a plan around rescuing, improving or transforming relationships, then the storyboard is an invaluable tool.

Principles are those self-evident and fundamental truths that are immutable and non-negotiable:

1   You can't be all things to all people.
2   Relationship approach is first a choice, then a responsibility.
3   Customers are the reason suppliers exist.
4   The right people, doing the right things, in the right way, at the right time, for the right reasons will deliver the best possible outcomes.
5   Insanity is doing the same things and expecting different results.
6   Relationship management is a process, not an event, and a journey, not a destination.

Principles are, or should be, universal and non-specific to any faith, culture, country or business sector, and hold true in all circumstances. The six 0 to 10RM principles are as much life principles as they are business relationship principles. You will instantly recognise them from your own experiences, both personal and business. They are, in effect, old sayings handed down from generation to generation. They survive because they are profound and timeless.

Each statement of principle in the storyboard is associated with a picture to support the words. Use the pictures as images to interpret the statement of principle. The six principles underpin the five themes from which the associated tools evolve. Tools within each of the themes are discussed in detail in the following chapters.

While we could start our journey anywhere on the storyboard, depending on the circumstance and the story being told, to give flow and a little structure we will start with Principle 1 and work from top to bottom in the simplified storyboard, engaging both the principles and themes. The complete illustrated storyboard is also available on the 0 to 10RM website at <www.0to10rm.com>.

Storyboard Part A—framework and components

Part A, Framework and components, comprises three principles and two themes. As the directions of the arrows indicate, the principles lead into the themes, and the principles underpin the detail associated with each theme.

Principle 1: You can't be all things to all people

No longer can a single organisation, whether in the public or private sector, be all things to all people. Confrontation is turning into cooperation, competition into collaboration, and often conflicting strategies into shared vision and common goals. The business landscape is littered with poorly performing firms who have refused to share, exploit synergies and leverage core competencies, both internally and externally. So it is with relationships. There is no single relationship approach or performance level that fits all circumstances. The nature of the participants at an individual, group and organisational level, the products and services involved, and the operating environment are all factors to be considered in improving relationships and performance.

It's all very well talking and engaging in external customer and supplier relationships, but internally we don't have our own act together. We can go no further until the quality and performance of our internal relationships improve. We haven't got the time to drain the swamp because we are up to our armpits in alligators. Working in these organisational silos we are so busy solving the day to day problems, fighting among ourselves and trying unsuccessfully to please everyone that we can't prioritise and get onto the big opportunities.

Operations and service manager—manufacturing sector

I have been involved with the development of strategic partnerships and alliances for more than two decades, and I have seen that two myths have evolved about these collaborative, trust and transparency based relationships. First, that they are easy and second, that they are suitable for everyone. While compelling in theory, not all business relationships can or should be partnerships and alliances. Organisations simply do not have the time or resources for developing partnerships and alliances exclusively, nor will there always be the right organisational alignment, or compelling strategic and commercial value for them to engage in such high-level relationships.

You can't be all things to all people, but you can be the right things to the right people. While the all-singing, all-dancing busker will appeal to some, the professional orchestra of skilled musicians is likely to appeal to a wider audience, be more adaptive and flexible, and deliver a better outcome to a higher standard. Sporting or team analogies are equally as good in helping us understand the practical application of Principle 1 (see figure 1.2). With rare exception, peak performance and achieving success in a complex work environment requires teamwork. In that regard, the whole is greater than the sum of the parts.

Figure 1.2: Principle 1—you can't be all things to all people

Principle 1 implies being fit for purpose. This requires clear objectives, roles and responsibilities; a willingness and capability to meet or exceed agreed requirements; flexibility to work independently or collaborate with others; the leadership, vision and courage to sometimes say 'no'; and a longer term, strategic perspective, as well as a shorter term operational view.

The reality is that you can't keep everyone happy all of the time. Try, and you will end up pleasing no-one. There is a decision point, based on diminishing returns, beyond which the cost of the effort, in both financial and non-financial terms, will outweigh the benefits gained. This dilemma often presents as suppliers not being able to say 'no' to customer requests or believing all business relationships need to be treated as equal. No two relationships are the same. Even politicians understand the impossibility of catering to all constituents. The one stop shop strategy will rarely satisfy all those people from the top to the bottom end of the market. Rarely, if ever, does one size fit all. In the area of organisational change, total employee engagement is very rare. There will always be, at the very least, a small minority for which the change is perceived as negative.

The alternative to Principle 1 — being all things to all people — often presents as a non-strategic, undifferentiated approach to all customers and suppliers, and creates misaligned and unproductive working relationships that waste time, resources and energy. Ironically this alternative approach implies that there is little choice when it comes to engaging customers and suppliers. In fact there are plenty of legitimate relationship choices. This is the thought that underpins Principle 2.

Question: Are you trying to be all things to all people? How does this behaviour apply to your own work environment, your life experiences and the relationships you are seeking to better understand?

Principle 2: Relationship approach is first a choice, then a responsibility

Being the right things to the right people means we have choices. There are plenty of choices around both the relationship approach we can take and the performance levels we can expect. But in making those choices we have to accept the associated responsibilities. For example, in choosing a collaborative approach based on trust and transparency, there are different responsibilities around sharing previously restricted information on innovation and new product development early and often. As it happens, Principle 2 (see figure 1.3) is a life principle, not just a business relationship principle. We all have choices around relationships, health, education and so on, but in making and taking those choices we need to understand the responsibilities and commitments required to turn each choice into reality. This is a two-edged sword: there will almost always be consequences in not honouring those responsibilities and commitments.

Figure 1.3: Principle 2 — relationship approach is first a choice, then a responsibility

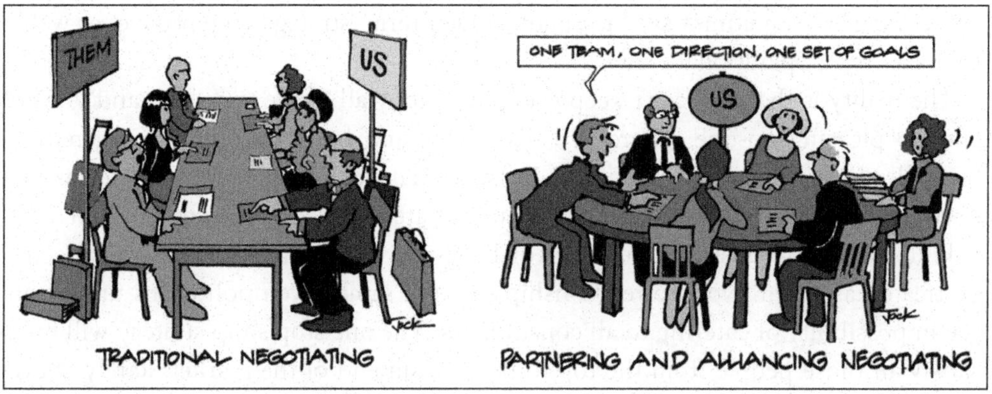
THEM
US
TRADITIONAL NEGOTIATING
ONE TEAM, ONE DIRECTION, ONE SET OF GOALS
US
PARTNERING AND ALLIANCING NEGOTIATING

I have been involved in negotiations where the negotiating parties were classic proponents at the square table, just like those in figure 1.3. In particular, in high level relationships the adversarial defend and protect, good guy–bad guy, them and us, win–lose approach that delivers poor performance needs to change—to an interdependent, win–win, collaborative, One Team partnering approach based on trust and transparency around common goals for mutual benefit. The responsibilities associated with this desired future state were fundamentally different and challenging for all the parties involved. These responsibilities and associated commitments affected the strategies developed; the choice of structures that supported the strategies; the competencies and qualities of the people engaged; and the relationship management and other business processes developed and implemented. Last, but by no means least, these responsibilities influenced the high performance culture that was required to sustain the value-adding strategies.

We have a high impact supplier relationship with this company, which is a virtual monopoly. They talk, negotiate, act and behave like a monopoly and are domineering, adversarial, controlling, dismissive and arrogant. So we match force with force and the result is most times lose–lose. We have no choice but to deal with them but there must be other ways to engage the relationship. There has to be a better way!

Chief information officer — IT sector

If Principle 2 holds true, what are the choices around the relationship approaches that can be taken and the performance levels that can be achieved? What do the associated responsibilities look like? These questions go to the heart of Theme 1—the 0 to 10RM Matrix (see figure 1.4, overleaf).

Question: What relationship choices and associated responsibilities have you made? What affect has this had on your own work environment, your life experiences and the relationships you are seeking to better understand?

Theme 1: 0 to 10RM Matrix—Understanding and mapping the current state and desired future state of your most important relationships

Figure 1.4 shows the 0 to 10RM Matrix, which is the framework or lens through which business relationships can be understood and managed. As in any aspect of life, it is difficult to see the full picture when you are sitting inside the frame. The 0 to 10RM Matrix allows us to sit outside the frame to take a strategic view of the relationship choices available, the associated responsibilities to be taken and the benefits or performance outcomes to be achieved.

In short, the 0 to 10RM Matrix identifies 11 legitimate relationship types and 11 performance levels. Relationship types could range from combative through to community within the vendor, supplier and partner segments. Performance levels could range from unsustainable through to satisfactory, world class or superior. Within this relationship type performance framework, we can understand and plot the current state and desired future state for the relationship, and develop and implement a journey management improvement plan to bridge the gap. We can also identify those secondary points, which appear as pockets or outliers of attitude, behaviour, practice or performance that differ in some way from the current state.

Figure 1.4: Theme 1 — the 0 to 10RM Matrix

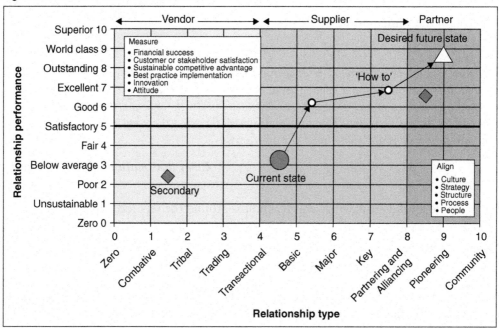

The five relationship components of culture, strategy, structure, process and people that make up each of the relationship types shown on the horizontal axis in figure 1.4 can be used to help understand the degree of alignment between the relationship parties. Six key results areas (KRAs) are associated with the relationship performance scale, which is shown on the vertical axis in figure 1.4, and these are used to measure relationship outcomes or success. The six KRAs are: financial success, customer–stakeholder satisfaction, sustainable competitive advantage, best practice implementation, innovation and attitude.

We have a critically important relationship but it is currently in the ditch. Very adversarial, lots of firefighting around problems and complaints. We have to either exit/terminate the relationship or take it to a completely different level, a new paradigm. But where to and how?

Chief procurement officer — banking and finance sector

The 0 to 10RM Matrix is the core diagnostic tool by which one or all parties in the relationship can engage in an open, honest, factual, no-blame discussion. From these internal and external conversations the relationship parties have the potential to reach a common understanding and alignment on the relationship improvement strategy.

The Relationship Alignment Diagnostic (RAD) (see figure 1.5, overleaf) and the Relationship Strategy Map (see figure 1.6, overleaf) are logical tools that arise from the 0 to 10RM Matrix. The RAD is a relationship health check or diagnostic that identifies and aligns each party's relationship approach to current and future performance. This allows a roadmap strategy or action plan to be developed to bridge the gap between the two prime anchor points for the relationship — the current state and the desired future state.

The 0 to 10RM Relationship Strategy Map enables the strategic analysis of many relationships across external market segments, internal business units and functions, product and service lines, and the development of relationship improvement strategies directly supporting the organisation's broader business goals and objectives.

The practical application of the 0 to 10RM Matrix, relationship alignment diagnostic and the relationship strategy map is discussed further in chapter 2.

While some organisations have trouble with key customers and suppliers, we have challenges with environmental groups, regulators and government departments all of whom have the potential to negatively impact on our license to operate. These are critical relationships; what can we do to improve them?

CEO — utilities sector

Principle 3: Customers are the reason suppliers exist

Most of us have heard of the wry saying 'If you don't have a customer, you don't have a job'. This applies internally or externally to the organisation. In that regard, customers are truly the reason suppliers exist. This principle does not mean customers (internal or external) are always right or that 'We need to keep the customer happy at any cost', but customers are the driving force behind implementing business strategy, and delivering shareholder and stakeholder value.

Figure 1.5: 0 to 10RM relationship alignment diagnostic (RAD)

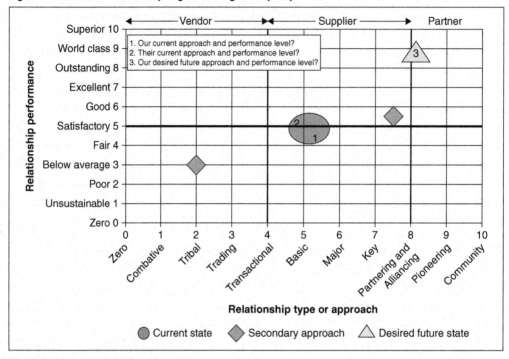

Figure 1.6: 0 to 10RM relationship strategy map

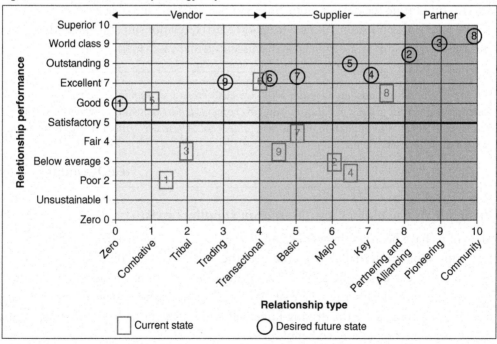

That Principle 3 is so self-evident makes it all the more surprising that only now, after the 2008–09 global financial crisis, are financial institutions required to act in the best interests of the client or customer. Even more surprising is the resistance from some within the industry to enacting this fundamental change. The often referred to catchcry caveat emptor (buyer beware) is not a principle but a safety net, frequently used by unscrupulous suppliers to exploit their advantages over unsuspecting customers.

Although there are isolated examples of outstanding performance, normally involving special projects and particular people, the relationship at this point can be characterised by defensive and protective behaviour; over-promise and under-delivery on product development; poor service levels; stock outs at critical times; occasional quality problems; high staff turnover issues; variable project performance; unnecessary or unreasonable cost variations; price and margin pressure; poor communications and information sharing; significant rework and inefficient workarounds; duplication of activities and overall a complete lack of trust between the companies. It is indeed a remarkable relationship, but for all the wrong reasons! There is a total lack of customer focus. What can we do to implement a step change?

Key account manager — energy, oil and gas sector

Good suppliers are entitled to challenge with respect, especially if a win–win option for mutual benefit is involved. As equal partners, both customer and supplier get to sit at the front of the truck (or around the same table), each with different roles and responsibilities (see figure 1.7, overleaf). Customers can be viewed in the broadest possible context as recipients of products and services. Principle 3 implies that the flow of value is from supplier to customer, who in turn becomes the supplier in the next step of the value or supply chain. Working together cooperatively and collaboratively will create abundance rather than scarcity.

As shown in figure 1.7, Principle 3 does not ignore the complexity involved in meeting or exceeding requirements, or the pressures and demands from customers, stakeholders and globalisation generally. Also implicit in Principle 3 is the importance of understanding each other's individual strategies, business drivers and the value propositions that underpin the relationship and the relationship approach taken. Value propositions are those opportunities or benefits to be gained beyond just a cheap price or a low cost. This is why making the right choices up front and understanding each party's willingness and capability are so critical.

At a deeper level, this principle is linked to a higher calling: that of being in the service of others. The service or act of serving has its own inherent rewards. This is particularly relevant to community relationships, not-for-profit activities, social and environmental issues and the growing number of organisations that are conscious of their environmental and social responsibilities. Ignore or compromise Principle 3 at your peril.

Question: Can you imagine a world without customers? How do customers influence your own work environment, your life experiences and the relationships you are seeking to better understand?

Figure 1.7: Principle 3—customers are the reason suppliers exist

Theme 2: Key Components—Understanding the key components that make business relationships work

We know where we are but we have no collective agreement on where we are going. Without strong leadership, clear roles and responsibilities, documented and agreed improvement goals and SMART (specific, measurable, achievable, relevant, trackable) metrics, the wheels are spinning but there is no traction to relationship and business improvement.

Project manager—construction sector

Fifty per cent of our revenue is with two accounts, both of which are poorly performing transactional relationships. Our future is at risk unless we can secure a more sustainable, higher performing, strategic engagement. We don't fully understand the underpinning value propositions and the leadership is failing. What can we do?

Group general manager—chemical sector

The seven key components provide an insight into the detail that underpins high performance relationships, wherever they may lie on the 0 to 10RM Matrix. They provide the link between the strategy of the organisation and the delivery of results. The key components are:

- Leadership and people who are willing and capable to lead and make the relationship work.

- Value propositions that are those benefits and opportunities supported by executive commitment that underpin the relationship beyond a cheap price or low cost.

- Legal contracts and agreements to document legal and binding obligations of the relationship, such as scope, requirements, prices and costs, and terms and conditions.

- The relationship charter to document the vision and mission, key objectives and guiding principles that are the basis of the moral agreement between the relationship parties.

- Key performance indicator (KPI) relationship and performance scorecards, that is the metrics, to form the basis for aligning performance measurement, risk management, relationship management, behaviours, continuous improvement and remuneration.

- Risk–reward models to build trust and transparency, and allow both risk and reward associated with the relationship to be shared and fairly managed.

- Relationship business plans, strategy or action plans to provide the roadmap for relationship improvement.

- Governance and leadership to define the stewardship, structures and interfaces, roles and responsibilities for managing the relationship(s).

The devil is in the detail and Theme 2 (see figure 1.8, overleaf) is all about managing the detail. This is often where subject matter experts—such as financial, legal, technical, policy and planning, insurance, contract management staff, with the help of relationship facilitators—are required to help, often through the application of special purpose or high-impact workshops.

Each of the relationship key components and their practical application are discussed in greater detail in chapter 4.

Figure 1.8: Theme 2—understanding the key components that make business relationships work

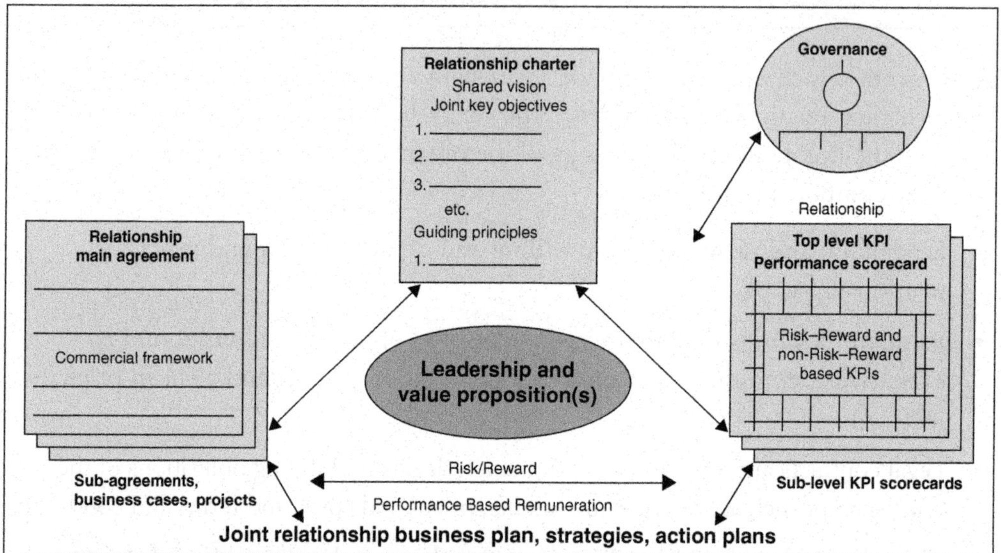

Key objectives	Initiatives	Who	By when
1.	1.1 1.2		
2.	2.1 2.2		
	etc.		

Storyboard Part B—people and change

People and change are central elements in high performance relationship management. Part B of the storyboard comprises Theme 3 and Principle 4. This is the critical link between the 0 to 10RM framework and the components that go into understanding relationships, and the supporting complexity and the process around journey management that are the subject matter of part C of the storyboard. In short, relationships are all about people, with change being a critical constant in all our lives. People and change are inextricably linked. Effectively managing change is what successful people do in living fulfilling and rewarding lives.

Principle 4: The right people doing the right things, in the right way, at the right time for the right reasons will deliver the best possible outcomes

Principle 4 is all about rightness. It is predicated on good leaders developing the right strategy and driving the right value propositions for the relationship. Principle 4 is directly associated with the Bus of Change as seen in figure 1.9, which is one of the most enduring concepts within 0 to 10RM. Over the past 15 years it has proven to be an effective and non-threatening change management tool to engage a discussion around the different types of people, attitudes and behaviours that live in relationships. Understanding who are the innovators (3 per cent), who are the early adaptors (13 per cent), who are the followers (68 per cent) and who are the saboteurs (16 per cent of the population) will be critical to the effective management of change. The Bus of Change is symbolic of the complexities of human behaviour that appear along the change journey.

Figure 1.9: Principle 4—the right people doing the right things, in the right way, at the right time, for the right reasons will deliver the best possible outcomes

High performance relationships are fundamentally about people and the effective management of change. People display different levels of adaptability to change. Some fear it; some are in denial of it; some undermine it, while others openly embrace change. I can think of no successful organisation or transformational initiative where people have not been key to the effective management of change.

There is a direct impact on personal development; employee engagement, loyalty, retention; and the way individuals behave and act. We are able to mobilise the people involved in Principle 4 and on the Bus of Change through the Let's Go theme (figure 1.10). This mobilisation enables the effective engagement of stakeholders in the common understanding, language and practice of relationship management and the associated change processes.

We've got some people acting like dinosaurs in the organisation, who still think that relationships are for family, friends and pets. They have opponents not colleagues and competitors not peers. They are part of the problem not the solution. Yet others are absolute champions, real leaders who are totally committed. How do we harness the power of people and their potential?

HR manager — education sector

The ability of people to cope with change will determine both the quality and speed of improvement. If the focus is placed on the innovators, the saboteurs who overtly or covertly act to undermine the change process can be managed up or out, or relocated to a place where they do no harm. In the context of 0 to 10RM, innovators and saboteurs share three common qualities. Both groups are stubborn, passionate and unreasonable people. Innovators use these qualities constructively up the front of the bus, and saboteurs use these qualities destructively down the back of the bus. Indeed one of the greatest achievements in any relationship is to turn the self-confessed saboteur into a relationship champion. The positive impact on relationships, communications and productivity of achieving this is quite remarkable.

Question: Where do you see yourself sitting on the bus? Are you an innovator, early adaptor, follower or saboteur when it comes to your own work environment and the relationships you are seeking to better understand?

Theme 3: The Let's Go change model — how to involve the right people and effectively manage change

The management and leadership of people and change can be complex and challenging, as anyone involved in a large organisational restructure will testify. The Let's Go model (see figure 1.10) is designed to be simple and effective in generating a common understanding, language and practice at all levels around change management and relationship improvement.

Figure 1.10: Theme 3—the Let's Go change model, or how to involve the right people and effectively manage change

Mobilising the Bus of Change and the people inside the bus on a journey of improvement from the current state to the desired future state can be achieved by using the Let's Go change model. It is a call to action. This theme brings to the fore the importance of people and effectively managing change together at both the strategic and tactical level. The model can be used to build Let's Go action plans using the simple Five Ws. These are:

- **Where** are we now and where are we going to?
- **Why** are we going there?
- **What** do we have to do to get there?
- **Who** is involved?
- **When** will we start and finish and when are the milestones?

These Let's Go action plans effectively engage and empower people and teams on the improvement journey in a structured and sustainable way, which gains commitment and develops ownership for the next steps.

The Let's Go change model enables people at all levels to tell stories, develop clarity around their roles and responsibilities, share learnings, and engage constructively and positively in the relationship improvement process. The investment at this point is in time and people. The roles of relationship sponsors, managers and facilitators as change champions turn out to be a key ingredient for driving transformational change within the organisation. Fundamental shifts in attitudes, behaviours, work practices and business outcomes are often the result. This in turn has a significant impact on the development of a high performance culture, the discovery of new paradigms and the delivery of sustainable competitive advantage.

The champions have left — one promoted and the other has gone to a CEO role with another company and we've lost our way. We spent two years entering the new relationship paradigm and it took six months to revert to the old ways. We need to reinvent, re-energise and re-set the relationship, and get it back on track, but how?

Marketing and sales manager — entertainment sector

All the elements that make up the Let's Go theme image (figure 1.10 on page 19) and the Bus of Change (figure 1.9 on page 17), including the role of high performance cultures and the relationship managers and relationship facilitators as change champions, are discussed in detail in chapter 5.

Storyboard Part C — journey management

Part C of the storyboard comprises themes 4 and 5 and principles 5 and 6. They involve the elements that need to be present for the journey from the current state to the desired future state of relationships. We will look at the phases that the journey is likely to take from trust building to innovation to paradigm shifting. Understanding what this phase change feels like, what the motivators and process steps are in delivering desired outcomes underpins part C.

Principle 5: Insanity is doing the same things and expecting different results

Change in people and relationships and paradigms are inextricably linked. There are shifters, pioneers and followers of new paradigms. Sometimes, ironically, it is the leaders in the organisation who don't see or even support the new paradigms because they are invested in the prevailing paradigms.

If you always do what you have always done, you will always get what you have always got. In moving from a current state to a desired future state, other than for factors outside our control, we need to do something different to get there. Change can look like doing better, at a stretch, or doing things differently and delivering breakthroughs. Principle 5 pictured in figure 1.11 aims to provoke us to challenge the status quo and the prevailing paradigms, to seek out new alternatives. 'But we have always done it this way' and 'This is the way we do business around here' are statements of attitude and behaviour fundamentally challenged by this principle.

In not doing things differently, but expecting different results, we often become victims, lacking control over our own destiny. We become more reactive than proactive. Principle 5 challenges us to act and rise above our current state, and it infers that we can all make a difference, individually and collectively. This applies at a personal, business or global level. Losing weight to achieve better health, implementing a new business process to improve productivity or reducing home energy consumption to positively affect climate change are all within our capability to make a difference.

Figure 1.11: Principle 5—insanity is doing the same things and expecting different results

They defend and protect everything, tell us nothing until the last minute, talk partnering and expect us to react. The talk is of collaboration but the old attitudes and behaviours have not changed. Then there is finger pointing and blame when we can't react quickly enough or respond effectively to meet their requirements In Full On Time, to A1 specification.

Sales and service manager — government sector

Innovation is a key objective for many relationships. It not only requires doing things better but also doing things differently. Innovation involves both stretch and breakthrough improvement, and challenging the status quo. Principle 5, illustrated in figure 1.11 with the logs being dragged and rolled to signify stretch and breakthrough improvement, becomes the basis for challenging existing norms and the status quo. It is sometimes a fact that you need to feel the pain and frustration of pulling the log before you can understand that there is a different way. But do we need to reach a point of crisis or breaking point before the need to change and improve becomes an urgent imperative? Why can't those compelling alternative options, or new paradigms, be engaged before the crisis point is reached?

Improvements such as implementing co-location, process and systems integration, joint product development, changes in leaders and leadership styles, improved development cycle times, supplier-based inventory management, joint training, secondments and people exchange across the organisations are just some examples that will be explored in later chapters.

Question: Are you pulling or rolling the logs? Are you working on stretch or breakthrough improvement in your own work environment and the relationships you are seeking to better understand?

Theme 4: The relationship development curve

What does the relationship journey look and feel like? The relationship development curve (see figure 1.12), puts the stretch and breakthrough improvement journey into perspective in terms of the timeline, phases and outcomes involved. Relationship development and improvement turn out to be a phased rollercoaster ride. This is often initiated by crisis or is introduced to prevent or pre-empt an impending crisis. Priority actions initially involve getting the basics right (products and services In Full On Time to A1 specification — IFOTA1); developing trust through performance improvement; delivering on commitments; and then looking for breakthroughs and new paradigms to prevent relapses and to continuously improve.

The crisis point often occurs when the question is posed as to whether to terminate the relationship or to fundamentally reset and rethink the approach and performance paradigms. At this juncture all parties involved will need to answer the value question, namely, 'What value is this relationship delivering for my organisation over the alternatives?'

Figure 1.12: Theme 4 — the relationship development curve

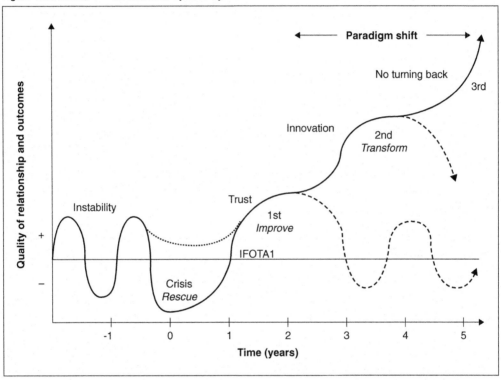

Knowing where you are on the relationship development curve, how you got there and what is taking the relationship forward are key questions in managing the journey of relationship rescue, relationship improvement or relationship transformation. The challenge for many organisations is how to accelerate the trajectory into the next phase. Typically, the most challenging times for any relationship, other than the initial point of crisis, is the top of each phase when the development curve starts to flatten out or plateau (see figure 1.12). Delivering on initial promises IFOTA1, and building trust and credibility by doing what we said we would are in large part stretching the existing paradigm. Moving from phase one to phase two requires genuine step change and breakthrough improvement, such as engaging open decision making processes and the joint sharing of information and risk, often involving previously classified data.

We have done a good job getting out of the quality and service crisis, but progress has stalled. We have islands of cooperation with sharks in between. The relationship drifts from one year to another. It is doing OK, but we just can't get it to the next level of engagement and performance. This is fine for now but is not sustainable in the longer term. The challenge is, how do we achieve the activation energy to accelerate up the next phase of improvement? How do we engage the new paradigms?

Procurement manager — healthcare sector

The journey of improvement is neither clear nor certain. Challenges and opportunities continually arise. Sustaining the relationship approach and performance beyond the life of key people; developing greater strength in terms of relationship champions to sustain the effort and workload; and maintaining executive support and commitment are ongoing challenges. This is especially the case with a continuing flow of new C people — CEOs, COOs, CFOs — and others, all wanting to make a difference, into and out of corner offices. Dealing with this aspect of change will be fundamental to effective journey management. The relationship development curve is discussed in greater detail in chapter 6.

Principle 6: Relationship management is a process, not an event, and a journey, not a destination

Just as we have key components to manage the relationship, we need a process to manage the journey. In that regard, all six 0 to 10RM principles and the five themes are linked. The process is always improving and the journey is continuous.

I think 99.9 per cent of people are decent, fair-minded and reasonable. So why is it that for the majority of the time we don't trust each other. We keep secrets, don't honour our commitments and the relationship is full of unpleasant surprises. A solution exists and we as one team have to find it. But how?

Business manager — utilities sector

Principle 6, supported by the picture of a bike rider representing the relationship navigating the long, winding and often rocky but enlightened road to improvement (see figure 1.13), is a visual reminder of the journey. It reinforces the key point that the journey takes time and will involve multiple steps. These steps would need to be safe, legal and logical, building varying degrees of trust, often around aligned or common goals for mutual benefit. The snipers or saboteurs from the

Bus of Change (figure 1.9 on page 17) also need to be managed, along with the innovators, early adaptors and followers.

Figure 1.13: Principle 6—relationship management is a process not an event, and a journey not a destination

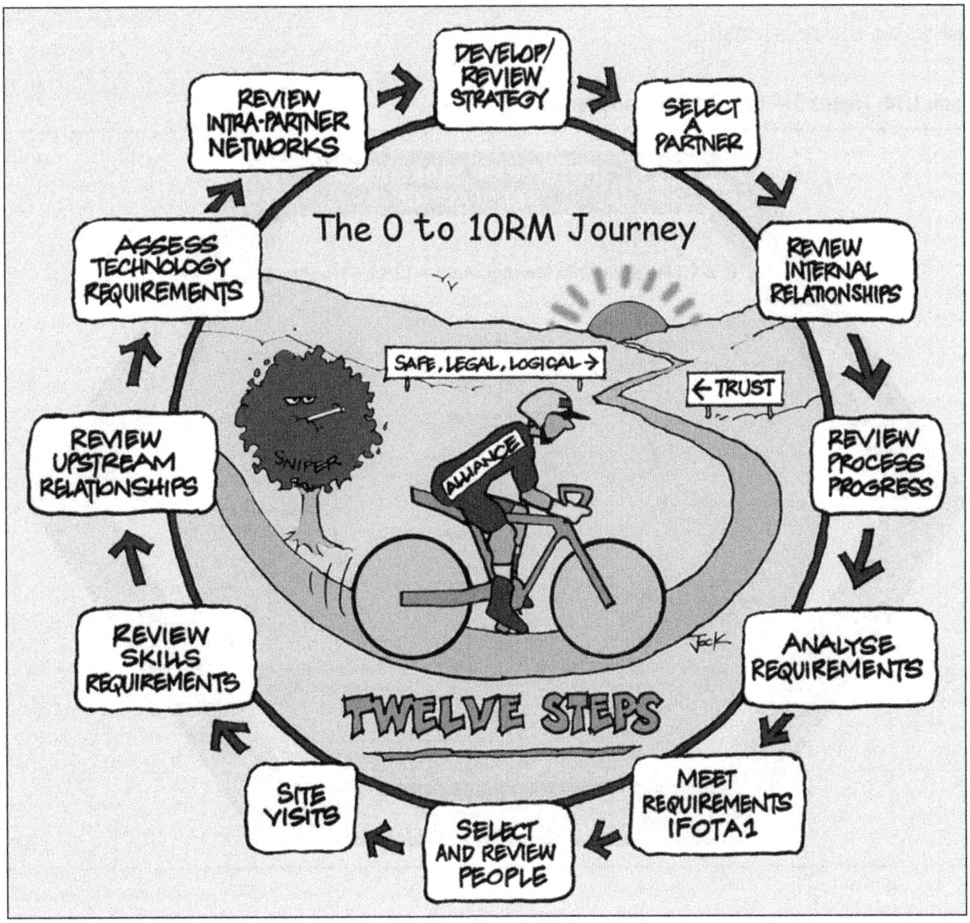

Question: What relationship journeys are you undertaking and what processes are you using to get there?

Theme 5: The 12/12/6 roadmap steps

In managing this journey into the 'how to' phase, there are at least 12 motivators to be monitored, 12 steps to be considered and six outcomes to be delivered on (12/12/6). This is Theme 5 (see figure 1.14, overleaf), comprising motivators, steps and outcomes. The 12/12/6 roadmap model provides the ideal balance between structure and flexibility in understanding the motivators (attitudes and behaviours) that drive the relationship forward, managing the steps to be implemented and delivering financial and non-financial mutually beneficial outcomes. Entire workshops, associated strategies and Let's Go action plans can be devoted to this one theme and image alone. The 12/12/6 roadmap model can be used many times to review and improve, re-ground, re-calibrate and re-set the relationship.

Figure 1.14: Theme 5—the 12/12/6 roadmap

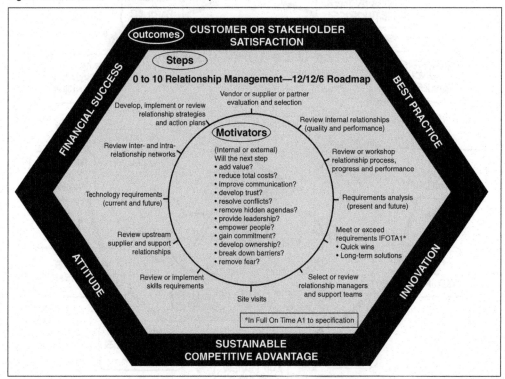

The 'plan, do, measure and improve' cycle associated with Theme 5 allows for the relationship improvement to be delivered and milestone points to be reached, and for the progress and results to be plotted on the 0 to 10RM Matrix. This brings us back to Theme 1 and the ability to loop back to the beginning of the storyboard to begin the next phase or cycle of relationship improvement.

We didn't engage, enable or inform enough people at the operating level early enough in the change and improvement process. That is, those people who make these partnerships and alliances work. We also didn't engage the legal and financial folk as critical stakeholders early enough. They are often the traditional custodians of corporate rules and policies. It's all very well having the folks in the Head Shed sign off on the Strategic Alliance agreement and the Partnering Relationship charter, but the teams and people on the ground, who have to make it all work, are still in the dark. What are the details around which this improvement journey works? How do we get people's buy-in and do it better the next time?

Business development and alliance manager — mining and resources sector

In this way the 0 to 10RM storyboard becomes a truly integrated and looped set of principles, models and tools allowing understanding and engagement at all levels of the organisation. The journey of continuous improvement can be effectively led and managed, its progress measured and the results used to demonstrate benchmarked and superior value for money outcomes. It can work for any relationship, internal or external to the organisation, in any country or any market sector.

The aspects surrounding the 12/12/6 roadmap, namely the motivators, steps and outcomes, are discussed in greater detail in chapter 7.

Summary

- Six principles and five themes underpin 0 to 10 Relationship Management (0 to 10RM) and, when combined into a single picture, they form the 0 to 10RM storyboard.

- The 0 to 10RM storyboard provides the basis upon which to take a journey; a framework to tell stories; a set of engagement tools to simplify complexity; and a way to proactively manage and improve business relationships.

- The six 0 to 10RM principles are universal, self-evident truths that provide a true north bearing for understanding relationships. They are:
  - You can't be all things to all people.
  - Relationship approach is first a choice and then a responsibility.
  - Customers are the reason suppliers exist.
  - The right people, doing the right things, in the right way, at the right time, for the right reasons will deliver the best possible outcomes.

- Insanity is doing the same things and expecting different results.
- Relationship management is a process, not an event, and a journey, not a destination.

- The five 0 to 10RM themes are the frameworks or models from which detailed understanding can be gained. They are the 0 to 10RM Matrix; the key components; the Let's Go change model; the relationship development curve; and the 12/12/6 journey management roadmap.

- The 0 to 10RM storyboard can be used to:
  - develop a common understanding, common language and common practice around high performance relationship management
  - explain the past, present and future status of selected relationships
  - tell stories and simplify complexity
  - instill passion and challenge the status quo
  - engage people at all levels and add value
  - identify opportunities for relationship improvement to bridge the gap between the current state(s) and the desired future state(s).

Part A

FRAMEWORK AND COMPONENTS OF 0 TO 10 RELATIONSHIP MANAGEMENT

The 0 to 10RM Matrix

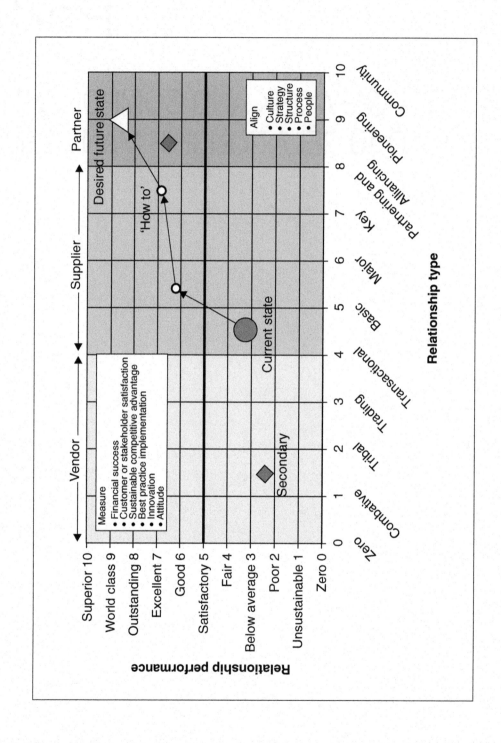

The 0 to 10RM Matrix (figure 1.4 on page 10) is the centrepiece of 0 to 10 Relationship Management. It presents 11 relationship types on the horizontal axis within three segments (vendor, supplier and partner), and 11 performance levels on the vertical axis. Within the matrix there are the two relationship anchor points — the current state and desired future state — and the 'how to' plan to bridge the gap between the two. Secondary states or relationship approaches sometimes present themselves, for better or worse, as pockets or outliers of attitude, behaviour, practice or performance that differ from the current state.

Means and ends are convertible terms in my philosophy of life ... and there is just the same inviolable connection between the means and the end, as there is between the seed and the tree.

Mahatma Gandhi[1]

The relationship approach is a means to an end, and not the end in itself. The end point or major journey milestones involve the delivery of the agreed business goals and objectives that underpin the journey. In that regard, relationship type or approach and performance are inextricably linked, in the same way as the seed and the tree. The means represents the process and the journey. It is the journey taken in bridging the gap between the current state and desired future state that provides the learnings and experience that lead to sustainability and improvement on subsequent journeys.

Unpacking the 0 to 10RM Matrix

The 0 to 10RM Matrix provides a framework — or a window or lens — through which all business relationships can be understood, managed and improved.

It has both strategic and tactical application, and it can be applied to internal or external relationships, upstream with suppliers or downstream with customers, to profit or non-profit organisations anywhere in the public or private sector.

First, we need to understand in detail the elements that make up the 0 to 10RM Matrix and how they all fit together. We can then interpret and relate the model to our own experiences.

The 0 to 10RM Matrix will provide a better understanding of how the world works in terms of business relationships. It has universal application across all relationships in all aspects of business and private life. 0 to 10RM is as relevant at home as it is at work. Relationships with builders, real estate agents, car dealers, school sports teams, retailers, tradespeople, professionals (doctors, lawyers, accountants), travel agents, friends and family, community groups, and buyers and sellers on the internet are all a part of our daily life.

For our organisation the 80:20 rule applies. That is 80 per cent of our revenue is generated with 20 per cent of our customers, but we have no effective customer or supplier segmentation tool. Put simply, for our most important relationships we don't understand where we are now, where we need to go to and why.

National sales manager — telecommunications sector

Chapter 1 presented Principles 1 and 2: you can't be all things to all people, and relationship approach is first a choice, then a responsibility. So what are the choices? The 0 to 10RM Matrix allows us to apply these principles and provides the relationship choices around which responsibilities can be taken. We can then be the right things to the right people.

Before we move forward on the 0 to 10RM journey we need to understand the parts that make up the whole of the 0 to 10RM Matrix. We can then overlay our experiences and stories to validate the model, then apply the model as a practical tool to start the journey of improvement to our desired future state(s). The six elements that make up the 0 to 10RM Matrix are:

1 relationship type scale (0 to 10)
2 relationship performance scale (0 to 10)
3 the two anchor points:
    - current state for the relationship
    - desired future state for the relationship
4 secondary state(s)

5   the green line
6   the 'how to' plan, which has milestones for bridging the gap between the current
    state and desired future state.

Relationship types in brief

Types of relationships are marked on the horizontal axis of the 0 to 10RM, which presents the 10 legitimate relationship types split into three categories or segments (vendor, supplier and partner), plus the zero relationship type. A short description of each of the 11 relationship types follows, and they are discussed in detail in chapter 3.

Summary of the 11 relationship types

*   *Type 0 Zero relationships* refers to a relationship approach where the choice is made, deliberately and for good reason(s), not to have a relationship with the customer, supplier, stakeholder or competitor in question. This choice could reflect misalignment or concerns about values, strategies, products and services. This kind of relationship might also characterise a new business opportunity, or a relationship that has been lost and is desired to be regained. A Zero relationship type can result from an exit strategy from an existing relationship.

*   *Type 1 Combative relationships* are confrontational, adversarial, aggressive, coercive, hostile, abrasive, divisive, uncooperative relationships. They are characterised by mistrust, win–lose, master–slave, bullying, control– compliance mentalities, as well as arrogance, the need for secrecy and a short-term profit focus. Combative relationships are typically conducted on the offensive, involve transfer of risk, and are often associated with hard- nosed, hard-dollar, tightly managed, detailed, one-sided contracts or service level agreements (SLAs).

*   *Type 2 Tribal relationships* are driven by self-interest. They are defensive and protective of information, profits, margins, costs, work practices, and the organisation and its departments, functions, position or power base. They are territorial, parochial, secretive, inward looking, often risk-adverse, relationships and resistant to change — they have many internal and external demarcations. People operating in these relationships are suspicious and mistrusting of others. The focus in achieving desired results often involves apportioning blame and finger pointing, rather than seeking mutually beneficial solutions.

- *Type 3 Trading relationships* live in a world of short-term opportunism, bargaining, bartering, contra arrangements, horse-trading, little loyalty, low margins and little differentiation. Products and services are seen as or treated as commodities. Trading relationships frequently involve formal or informal workarounds or networks that emphasise negotiating or doing the deal, retaining existing business, and getting or giving the purchase order at the best or lowest price.

- *Type 4 Transactional relationships* do business predominantly over the counter (electronic or retail), over the telephone, by fax or over the internet, with little or no negotiation involved. These relationships are based on standard terms and conditions. Driven by low cost, speed of transaction, richness and reach of information and service, they are often mechanical or robotic in nature, conducted at arms length, faceless, impersonal, automated, systems or technology-based relationships.

- *Type 5 Basic relationships* aim to *do and charge*. They are based on an agreed scope of work with a focus on price, quality and measured delivery of agreed requirements (products and services) In Full On Time to A1 specification (IFOTA1). This is often done through inflexible, prescriptive contracts or SLAs, won through referral, or a tender or competitive bid process. The focus is on the short- to medium-term and cost performance against budgets. Basic relationships are independent, often reactive and task driven. They have little focus on, or requirement for, innovation or continuous improvement, other than according to general market trends.

- *Type 6 Major relationships* aim to *do and improve* against agreed baselines. They focus on total systems cost reductions, over and above the In Full On Time to A1 specification delivery of agreed requirements. There is a sharing of business goals, objectives, performance drivers or measures, relevant financial and non-financial information, and opportunities for improvement. Major relationships are results driven, customer focused, continuous improvement and total quality-based relationships. There is a medium- to long-term focus on reducing or improving total costs rather than on adding value.

- *Type 7 Key relationships* aim to *do and add value* around agreed strategies. They are long-term, strategic relationships that exploit synergies between customer(s) and supplier(s). All relevant information is shared to minimise areas of conflict and promote high levels of cooperation and innovation, with the aim to add value and reduce total costs. This is in addition to meeting or exceeding a complex set of agreed requirements IFOTA1 specification. Key relationships are outcomes focused, proactive solutions selling, independent, preferred-supplier based, win–win relationships with multi-level interfaces.

- *Type 8 Partnering and Alliancing relationships* are based on trust and transparency around a shared vision, common goals and joint strategy for mutual benefit. They are collaborative, principle centred, interdependent, performance-driven relationships that have integrated interfaces, work teams and processes. Partnering and Alliancing relationships live in a world of seamless boundaries, frictionless commerce, shared risk–reward, performance-based remuneration and joint benchmarking. There is a shared governance of the relationship by joint leadership, management or operational teams holding themselves mutually accountable for the wellbeing and success of the relationship. Partnering and Alliancing relationships leverage core competencies for continuous and breakthrough improvement around a strategic scorecard of leading and lagging performance measures. These relationships are not only strategic but are also seen as critical to the long-term wellbeing and success of the partners. They are often viewed as relationship role models and used as vehicles for internal transformation.

- *Type 9 Pioneering relationships* capture the paradigm shifters and pioneers who dare to seek new relationship boundaries and break old rules. These are brave, bold and different relationships that have empowered, accountable leaders and teams comprising people who are passionate, stubborn and often unreasonable in their expectations. They find new solutions to seemingly impossible problems. Breakthrough thinking coupled with intelligent risk taking is encouraged, delivering both continuous and breakthrough improvement. Some examples of these improvements include transformational initiatives from the global to enterprise level, integrated project breakthrough teams, advanced multi-partner alliances, virtual enterprises, competitor collaboration, complex public–private sector relationships.

- *Type 10 Community relationships* include integrated supply chains and value chains, the extended enterprise, business cooperatives, internal networks and communities of practice, social enterprises, open source communities, social media, and community groups and organisations. The community is legacy building and collaborative. It delivers sustainable triple bottom line benefits (business, social and environmental) for a common purpose, around common goals for the common good. Communities are often multi-layered and complex. They can operate both online and offline, internally or externally, in the public and private sectors, for profit or not for profit. Participants share a common sense of community and working democracy, sharing information, leveraging skills and competencies, pooling and sharing resources across the community. Community relationships are populated by selfless, inclusive, giving and caring people who are open and outward looking, and linked by an interconnected and shared destiny.

The five elements of relationships

Relationship type is a function of five elements that need to be aligned by the relationship parties. These elements often present themselves as the mindsets, behaviours and practices that are deployed in the relationship. The five elements are:

- *Culture* — 'the way we do things around here', which arises from deep-rooted values and belief systems of the organisation.

- *Strategy* — how the business relates to its external environment.

- *Structure* — reflects the interfaces, the nature of the multi-level contacts, relationship governance and interaction between the parties in the relationship.

- *Process* — the motivators, steps or activities and outcomes involved in managing and improving the relationship.

- *People* — the relationship sponsors, relationship managers, relationship facilitators and those others involved directly in the relationship and their skills, competencies, roles and responsibilities; they include champions, pioneers, the passionate ones and other key influencers in the relationship.

Figure 2.1 summarises the make-up of the vendor, supplier and partner segments within the 0 to 10RM Matrix as seen through the five elements. Figure 2.1 also identifies a number of other factors such as trust, transparency, loyalty and their increasing involvement and relevance in relationship types from vendor, supplier through to partner segments.

The vendor, supplier and partner segments

Understanding the degree of alignment between customers and suppliers (see figure 2.1) in the five relationship elements of culture, strategy, structure, process and people across the 0 to 10 relationship types is critical to the management of all important relationships, not just partnerships and alliances.

There are two myths about Partnering and Alliancing relationships. First, that they are easy and, second, that they are for everyone. It is unlikely that any one company will engage in vendor or supplier or partner segment relationships exclusively. In reality, especially with larger organisations, a spread of vendor, supplier and partner relationships are required. The question is then, 'How can you have a mixture of cultures, strategies, structures, processes and people, within the one organisation, that can effectively manage such a wide range of relationships?' The answer is that organisations capable of higher level relationships are able to adapt and modify their operations and behaviour to suit lower level relationships.

Figure 2.1: vendor, supplier and partner segments

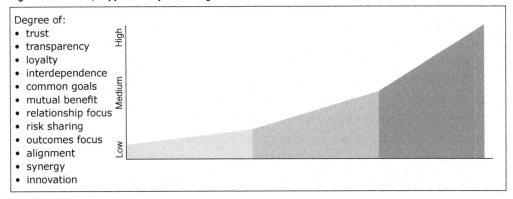

	Vendor	Supplier	Partner
Culture	Deal-making; cost and price focused; self-interest	Customer; continuous improvement; quality focused; common interest	Highly flexible; empowering; innovative; principle centred; common purpose
Strategy	Least cost or lowest price	Differentiation (cost or value); problem fixing; solution selling	Best and next practice; value-adding
Structure	Traditional interfaces; competitive tension	Multi-level interfaces; push–pull tension	Team-based interfaces; creative tension
Process	Buy and sell focus; coordination	Supply and procurement focus; cooperation	Shared ideas, and risk and reward; collaboration
People	Hard-nosed contract managers, sales or service representatives, traders, deal makers	Basic, major and key account managers and project managers; relationship facilitators	Partnering and Alliancing managers; pioneers coaches, community leaders

An organisation does not have to take on multiple personalities to effectively service this wide range of customers and suppliers. Organisations that have a base of supplier segment relationships will adjust their culture, strategy, structure, process and people to manage vendor relationships as required. Likewise, organisations with true partner segment relationship management capability have the flexibility to adapt themselves easily and effectively to managing relationships with both supplier and vendor segments.

The reverse scenario, however, does not hold. Organisations that have predominantly vendor segment characteristics, and no strategic intent to evolve, have difficulty developing and managing high-level supplier segment relationships and partner segment relationships. Likewise, Partnering and Alliancing relationships should not operate exclusively as experimental satellites distant from the rest of the business, as the broader benefits of collaboration, trust and transparency may be lost. The organisational equivalent of islands of cooperation, with sharks in between, is not an environment for sustaining high performance relationships.

There is often a cumulative effect in improving relationships within the vendor segment, then transferring the skills and knowledge learnt to the supplier segment and then the partner segment. Behaviours evolve that act as building blocks for each higher level relationship type. An example is changing culture from a focus on self-interest (vendor segment) to a focus on common interest (supplier segment) to a common purpose (partner segment). Relationship manager competencies will build from the sales representatives and traders operating Type 3 Trading relationships through to the competencies associated with Type 5 Basic, Type 6 Major and Type 7 Key relationships among account managers and onto the Type 8 Partnering and Alliancing (T8) relationship managers. Within the supplier segment, in particular, there will be an add-on effect in moving from Type 5 Basic *do and charge*, to Type 6 Major *do and improve*, to Type 7 Key *do and add value* relationships.

Relationship performance

Relationship performance is marked on the vertical axis of the matrix. There are 11 (0 to 10) possible performance levels. This axis on the matrix is about results, effectiveness and performance, and the impact of the deployment of those mindsets, behaviours and practices associated with the horizontal axis relationship types. The 11 performance levels are summarised here.

Summary of the 11 performance levels

0   *Zero*. This is normally the starting point for the relationship. No results have as yet been achieved. The relationship may still be in the planning phase, or the planning phase may have been completed but the implementation is yet to start or deliver results.

1   *Unsustainable*. The relationship is (a) just beginning or (b) almost finished or (c) on the verge of collapse. Significant change is required for improvement or an exit strategy needs to be developed or implemented. The relationship cannot be sustained at current perform ance levels because it is

completely failing to meet the organisations' expectations or requirements. This may be a function of internal and external factors.

2   *Poor.* The relationship performance is well below expectations in terms of meeting agreed requirements; in other words, it is miserable. This could be a function of an early days, immature relationship in which few results have been achieved, or unprofitable or uncompetitive performance levels against forecast expectations or requirements not being met.

3   *Below average.* This level of relationship performance will generally cause ordinary, and usually unacceptable results as determined against an agreed average benchmark(s) or target(s). This level of relationship performance should be a matter of concern or may possibly be a function of early days in the relationship development, though improvements are anticipated. Whatever the cause of this type of relationship performance, it is yet to deliver acceptable results. Current results are low service levels, product quality, profitability or growth, or are below break-even point, budget or forecast.

4   *Fair.* This level of relationship performance will produce a 'just okay or barely okay' result against expectations or requirements. There are no pleasant surprises at this level of performance. Only the minimum expected results or outcomes are likely to be achieved and there is unlikely to be profitable growth or sustainable competitive advantage. Results may be at or little better than break-even. This is mediocrity.

5   *Satisfactory.* This level of performance delivers an average, baseline result against requirements or expectations. It represents a midpoint and business as usual. Results will be nothing special. Organisations will not be happy, but neither will they be unhappy, with the results. Little or no excitement is involved at this level of performance. It is, however, often a position for change.

6   *Good.* Organisations are likely to be satisfied with this level of relationship performance. It produces a good, solid result against expectations or requirements, though ideally opportunities for improvement will be recognised. There will be conformance to requirements based on stretch targets. This is the first point of real acceptability on the results scale. Profitable or differentiated performance results are being achieved when benchmarked against competitors.

7   *Excellent! Wow! Great!* Excellent results, financial and non-financial leading and lagging, are likely to come from this level of relationship performance. Sustainable competitive advantage or profitable growth will be gained. There is room for

improvement only if additional benefits can be gained. At this performance level better than expected results are achieved against agreed requirements.

8 *Outstanding.* An outstanding level of relationship performance is likely to produce results that knock your socks off, or could be described as 'stand out' or 'best in class'. Organisations are likely to be delighted with the current performance or outcomes against expectations or requirements. These are likely to be benchmarked (internal or external) against competitive and sustainable results. Outstanding performance and the results it produces would be a very acceptable goal for most relationships.

9 *World class.* Results are likely to be described as top class; first class; or first division. Relationship performance is best in class or world class, as benchmarked against relevant KPIs. The relationship is a benchmark, role model and centre of excellence, and has a positive ripple effect on others. The relationship produces a self-sustaining momentum and creates a significant competitive advantage or profitable growth — the target for world class relationships. Once organisations achieve this level of performance they may need to focus on improving relationship type or resetting the performance bar.

10 *Superior.* Best of the best and top of the first division. This relationship performance is either unique or rare. It is as good as it gets in terms of performance and is the ultimate performance goal for any relationship type. Often it is more of a target than a reality. It is almost beyond measurement and offers the ultimate in competitive advantage. These performance results set the benchmarks for others to follow.

Key results areas

The level of performance achieved in the relationship (as marked on the vertical axis of figure 1.4 on page 10) will be a function of six performance criteria or key results areas (KRAs). These performance criteria, with selected examples of the metrics associated with them, are shown here. The six criteria are:

- *Financial success* is measured by such financial indicators as profitability, return on investment (ROI), total cost improvements, revenues, volumes, outputs, overall value for money, unit costs, transaction costs and working capital.

- *Customer or stakeholder satisfaction* is measured by such characteristics as quality, cost, schedule and service levels, response times, business case or project completion outcomes, survey results, flexibility and responsiveness.

- *Sustainable competitive advantage* is measured by such characteristics as market share or growth, customer and supplier loyalty and retention, amount of referred business, percentage of available business, percentage of negotiated business, and brand, image and reputation.

- *Best practice implementation* is measured by such characteristics as benchmarked efficiency, reliability, availability, capacity utilisation, implementation and application of systems, processes and procedures, degree of operational excellence.

- *Innovation* is measured by such characteristics as time to market, development cycle times, number and success rate of innovative ideas, continuous and breakthrough improvement, new paradigms engaged, innovation investment and allocation of resources.

- *Attitude* is measured by such characteristics as behaviours, mindsets, trust levels, leadership, communications, work practices, survey results, good news stories.

Each of these six performance criteria will be weighted differently for each type of relationship. For example, vendor relationships have a heavier financial performance weighting. As the relationship approach moves from Combative through to Community, there will also be a greater weighting or emphasis on the other five non-financial performance criteria.

The 0 to 10RM performance scale at first glance is subjective in description, using such terms as poor, fair, good, excellent. However, the performance scale can easily be linked to objective metrics, financial and non-financial, performance scorecards and associated KPIs, targets, and upper and lower performance limits. This, in turn, has a connection to the relationship charter goals and objectives, relationship value propositions and the business strategy.

The two anchor points

The two anchor points for the relationship are the current state and the desired future state, and these are a function of both the relationship approach and the performance level.

The current state is a representation of how each party approaches the relationship and the performance levels being achieved. The desired future state is the major milestone or end point to which one or all parties aspire. The desired future state is framed in the medium- to long-term in the context of the relationship under review.

To quote Steven Covey's second habit of highly successful people, 'Begin with the end in mind'.[2] Parties engaged in high performance relationships should begin their journey with the end points and major milestones in mind. This is about having strategic intent. The tactical aspects of the relationship journey on a daily, weekly or monthly basis will be composed of different and changing experiences, touch points and associations. However, the end points and major milestones will remain unchanged. Without understanding the two relationship anchor points, the journey of business improvement is rudderless.

Secondary approaches

Secondary approaches—pockets or outliers of attitudes, behaviour, practice and performance that are being displayed and delivered which, for better or worse, are different from the prime current state—are important. They can occur when the relationship is under pressure or stress, or when special projects or tasks are being implemented.

Positive secondary states may involve the practice or uptake of new paradigms, new practices or behaviours that the relationship as a whole is seeking to embed. Positive secondary states are where paradigm shifters and pioneers live. Interestingly, these folk can be anywhere within or outside the organisation, such as project teams practising joint problem solving for mutual benefit or open book costing and associated information sharing to accelerate project delivery. This is in contrast to the traditional closed book, non-transparent, fixed price or lump sum arrangement.

At other times negative secondary states indicate that pockets of the relationship have reverted to type: the dark side of the force is in operation, and hidden agendas, self-interested backroom deals or workarounds are in operation. Negative secondary states could also present as individuals or small groups overtly or covertly undermining collaborative initiatives and filtering or blocking the sharing of information that would speed up the process and benefits associated with collaboration. Unpleasant surprises, such as problems with quality, cost or schedule, are kept secret beyond a point where effective win–win solutions can be implemented. Commercial negotiations that turn combative can often undermine the good work done by others and negatively impact what would otherwise be a high performing, stable relationship.

Either way, it is critically important that positive and negative secondary states are understood and managed. They should be removed when their effect is negative and embraced when their impact on the relationship is positive. The reality is that, within complex, multi-site, multi-business, multi-partner relationships, there are likely to be a variety of secondary states—good, bad and indifferent.

The green line

The green line is the horizontal line at performance level 5 Satisfactory (see figure 1.4 on page 10). All relationships must strive to have performance at or ideally above the green line, irrespective of the relationship type engaged. In the context of the varying degrees of high and low performance, the green line is the defining and the dividing line between the two. Often a position for change, this is an average, midpoint, baseline result against requirements or expectations.

In interpreting the green line, consider the following personal analogy. Your partner or close friend is asked to rate your performance within the relationship across any and all criteria. 'Satisfactory' is their response. While you may not be devastated, you're unlikely to be flattered. Most people would see this as a point from which improvement can be made. So too in business, where typically only at the higher performance levels of good, excellent or beyond are relationships truly sustainable, contract terms extended, work scopes expanded and go to market re-bid options avoided.

The 'how to' plan

The 'how to' plan involves the strategies, initiatives, actions and timelines by which the gap between the current state and the desired future state is bridged, and the secondary approaches are effectively managed. The relationship management journey between the two anchor points is represented by the arrowed lines on figure 1.4 and will include milestones or interim goals, represented on the matrix as small circles on the 'how to' journey. While the intent is a continuous improvement journey with moments of breakthrough, the reality is that the journey is typically saw tooth in nature, particularly around performance, where the journey is often a two steps forward, then one step back affair. This is due to a variety of reasons, such as the difficulty of maintaining initial success, a settling effect in engaging and bringing the support teams up to speed, changing market conditions and under-estimating the impact of change on people and processes.

Even in the development of high performance Partnering and Alliancing relationships, it is the entire 0 to 10RM Matrix that needs to be understood and managed. Understanding the current state and the desired future state for your relationships and implementing the 'how to' plan to bridge the gap is the key to effective relationship management.

For some relationships it is simply a matter of improving performance and not moving to another relationship type — that is, doing what you are doing now, but better. In other relationships a paradigm shift is required to deliver the desired future state. Going from Type 7 Key to Type 8 Partnering and Alliancing on the relationship scale, or more generally from the vendor to partner segment, is not about incremental change. Nor is it about reducing costs by a few dollars; improving customer service a notch; getting stock levels down a smidgen; delivering products to market a few days earlier; or winning another percentage point of market share. Partnering and Alliancing relationships are about fundamental change and breakthrough improvements in attitude, mindset, practice, performance and behaviour.

Understanding the relationship types and performance levels required, up and down the supply chain as well as within the organisation, and aligning the individual relationship strategies with the corporate strategy will play a key role in delivering sustained competitive advantage and profitable growth.

My simple philosophy surrounding 0 to 10RM and life in general is encapsulated in what I call The Way:

The Way

- **Keep the faith** — in your guiding principles, values, beliefs.

- **Stay focused** — on your goals and objectives.

- **Enjoy the journey** — learn, have fun and celebrate success.

Business implications surrounding the 0 to 10RM Matrix

Understanding the current states of your relationships and agreeing on the desired future states have direct implications for all aspects of business. For example:

- *Business strategy and external relationships* — how the organisation goes to market and relates to its external environment, manages resources, utilises assets and spends its capital will depend on the anchor points; that is, the current state and desired future state.

- *Internal relationships (type and performance levels)* — this is the willingness and capability of internal relationships to meet or exceed customer, supplier and stakeholder expectations as expressed in the desired future state.

- *Customer and supplier evaluation and selection processes* — different relationship types require different criteria and processes around evaluation and selection. This will affect how sales and business development teams build and bid proposals.

- *Negotiation, leadership, communication and interpersonal styles* — these will differ significantly depending on relationship types and performance levels engaged.

- *Team dynamics, people and change management* — different challenges and opportunities will be presented at the people and change level depending on the relationship approach and performance level, current and future.

- *Commercial and contractual framework, and terms and conditions* — these will differ significantly depending on relationship types and involve varying skill sets, professions, special interest areas and subject matter expertise.

- *Business relationship manager competencies* — will vary across the relationship types, namely, skills, knowledge and attributes.

- *Relationship management and improvement strategies* — these will look different and be implemented differently.

- *Customer and supplier segmentation* — the 0 to 10RM Matrix is an ideal tool for segmenting customer and supplier relationships on agreed criteria.

Figure 2.2 is a simple representation of the relationship life cycle from initial strategy development and 'go to market' approach, through to the review and improvement cycle. Relationships and relationship management are a fundamental part of business strategy development and execution.

Figure 2.2 : relationship life cycle

How to determine which relationship approach is appropriate

If the relationship type and approach is the means to an end rather than an end in itself, how do you determine which relationship type or approach is appropriate to achieve the best possible business outcomes? For example, the determination of the appropriate relationship type is particularly relevant to the strategy and process taken in supplier evaluation and selection, and customer or supplier segmentation.

To enable effective selection or segmentation, a detailed analysis of the strategic value and commercial value associated with the products and services involved is required, together with an assessment of the willingness and capability of the relationship participants.[3] Chapter 7 considers a specific example of how this detailed analysis can be used in the process of evaluating and selecting a supplier partner.

Are some relationship types better than others?

How do we answer this question: 'It's all very well having a current state and desired future state within the 0 to 10RM framework, but this customer relationship is a complex relationship and all these relationship approaches operate at some time. How do we make sense of that in the context of the two anchor points and the improvement pathway between the two?'

The answer is that it is absolutely true that even for a single customer or supplier relationship all relationship types may be engaged at some point, especially in complex, multi-layered relationships. It would be wrong to suggest that one relationship type is better than another. As with different leadership styles, relationship approach will be specific to time, place and circumstance, in the short-, medium- or long-term. Good organisations need to understand how to deal with all the relationship types. In doing so, they will be adaptive to present and future challenges. However, there should always be an underpinning set of principles and value propositions that determine the prime relationship approach taken as both a current state and desired future state.

Take marriage as an analogy. The fact that we take vows — for better or worse, for richer or poorer, in sickness and in health, until death do us part — does not mean that we don't have to engage in daily mundane tasks, resolve the occasional disagreement or family dispute, deal with each other's self-interests, egos and idiosyncrasies, or take on broader community responsibilities. It comes down to the underlying principles, pledges or vows that underpin the relationship in the longer term.

In that regard marriage has similarities to a strategic Partnering or Alliancing relationship. That is, they are both principle centred, and based on trust and transparency around common goals for mutual benefit. Few alliances and partnerships travel so smoothly that they don't have to deal with varying levels of vendor and traditional supplier–customer behaviour. Deals are done; arms length, transaction-based systems are put in place; basic tasks implemented for delivery IFOTA1 specification. However, these activities are underpinned by a set of agreed, enduring, collaborative principles, and aligned, long-term, win–win goals. Yes, like a marriage, business relationships go off course from time to time. Overall, however, it is these core principles that set and reset the prime path of the relationship and underpin all the other relationship approaches that may be engaged, to whatever degree, for whatever reasons, from time to time. There are seven important takeaways around the 0 to 10RM Matrix:

- Relationships and their performance, internally and externally, are a fundamental part of business strategy. They will require an investment in time, people and money, and cannot operate in isolation.

- Achieving success is not about shifting all your customer and supplier relationships to Partnering and Alliancing relationships. Instead, you should optimise the relationship mix in both relationship type and performance level to best suit your business strategy. Remember Principle 1: You can't be all things to all people.

- It is not always the case that relationships need to move to the left or right on the horizontal relationship type scale. It may be the case that it is only performance levels that need to change.

- Know the two anchor points: the current state and the desired future state. Only from these two anchor points can the journey of improvement be managed. Start to think about the end you have in mind.

- Secondary points are important either as barriers to improvement to be removed or as role models to be emulated.

- At a minimum, all relationships must be above the green line (at or above number 5, satisfactory, on the performance scale).

- Organisations that have successful partner segment experience (Partnering and Alliancing, Pioneering, and Community relationships) typically have better performing vendor and supplier segment relationships.

Aligning culture, strategy, structure, process and people

Table 2.1 (overleaf), the 0 to 10RM relationship type characteristics matrix, provides descriptors of relationship culture, strategy, structure, process and people for the 10 relationship types, Combative through to Community, as indicated on the 0 to 10RM Matrix.

Table 2.1: the 0 to 10RM relationship type characteristics matrix

	T1 Combative	T2 Tribal	T3 Trading	T4 Transactional	T5 Basic	T6 Major	T7 Key	T8 Partnering and Alliancing	T9 Pioneering	T10 Community
5 Culture	Master–slave Adversarial Power, control and compliance Abrasive Taking culture	Defend and protect (internal or external) Inward focus Keep to themselves Risk averse Secretive	Opportunistic Deal doing and deal making Formal and informal workarounds and networks	Arms length Impersonal Service or transaction focus	Reactive Inputs focus Customer focus Get the job done	Outputs focus Quality and continuous Improvement-based values Do more for less	Outcomes Focus Value for money Open Proactive	Based on total trust and transparency Principle centred Value for money Collaborative	Who dares wins mindset and attitudes Calculated, educated, intelligent risk taking	Selfless Inclusive Giving Caring Open, outward-looking culture
1 Strategy	Offensive Win–lose profit focus Tough and hard-nosed contract focus	Defensive Protective Ownership focus Self interest Based on WIIFM	Deal based Limited loyalty Low margins and little or no differentiation Best of three quotes Commodity or price mindset	Technology driven Systems based Standard terms and conditions Low cost Speed and reach focus	*Do and charge* on agreed scope Tender based Cost plus Prescriptive IFOTA1 focus	*Do and Improve* against agreed baselines Best of breed Outsourcing IFOTA1 focus on reducing cost base	*Do and Add Value* around agreed strategies Exploiting synergies Preferred supplier Solution selling	Mutual benefit Interdependent Win–win Joint strategy Shared risk and reward Common goals	Brave, bold and different Leveraging core competencies Paradigm shifting	Triple bottom line and legacy focus Common purpose Common goals Common good Trust-based joint business plan
2 Structure	Bureaucratic Hierarchical Hostile interfaces and abrasive rub points	Parochial Silos, fiefdoms Clans, tribes, factions Territorial Interfaces & boundaries	Simple or single point of contact interfaces Face to face or electronic	Electronic or single point of contact	Simple, single or limited points of contact Face to face or electronic	Medium level multi-contact and contract management interfaces	Complex multi-level contract & contact management interfaces	Flat, team based, integrated interfaces Seamless boundaries	Empowered, flat and modular teams and interfaces	Extended supply chain and community interfaces Fuzzy Open source type structures

Process **4**	Legalistic Tightly managed one-sided contracts Risk transfer	Protect territory, information and knowledge, position and power base Many demarcations and hidden agendas	Trackable, traceable, deal based Efficiency and effectiveness focus Bargaining, bartering	Systems driven and automated Depersonalised Standard terms and conditions rule	Work to rule or Standard Basic account management IFOTA1 Focus	Major account plans linked to KPI measures and contractual obligations Detailed supply chain analysis or benchmarking	Strategic Key account plan leads relationship development & performance Regular BRADs Compatible processes	Joint business plan Flexible and adaptive performance and improvement based Joint ownership and governance Integrated processes	Best practice forums Stretch and breakthrough innovation and improvement processes	Health of community and legacy focus Personal Integrated or modular Few contracts
People **3**	Aggressive Confrontational Untrustworthy Arrogant, hostile, coercive communicators	Self-interest focus Us and them Protective and defensive Tribal loyalty	Short-term deal focus Work hard, play hard negotiators Deal makers, traders	Transaction or task driven Service oriented Technology driven or focused	Task driven Reactive relationship management skills	Results driven Focus on transfer of non-core competencies from customer to supplier	Outcomes driven Professional key account managers High accountability	Performance driven Principled Passionate Professional Fair minded and reasonable	Passionate Proud, stubborn and unreasonable in their expectations Pioneers and trailblazers	Selfless Giving, caring Working for community benefit and common good

Notes: BRADs = business review and development meetings; IFOTA1 = In Full On Time to A1 specification; WIIFM = what's in it for me

Experience suggests that there is a lead and lag effect associated with the five relationship alignment characteristics. Typically strategy (1) leads; structure (2) supports the strategy; good people (3) are embedded into the structure; the people develop and put in place roadmap processes (4) for managing the relationship journey; all of which ultimately impact on the culture (5). These ascending numbers 1 to 5 are listed on the characteristics matrix (table 2.1).

This information will be most helpful in understanding where the two anchor points are on the 0 to 10RM Matrix (that is, current state and desired future state of the relationship) and assist in answering the questions around what to stop, what to start and what to keep doing in order to achieve those end goals.

0 to 10 relationship alignment diagnostic

How is a relationship business model like the 0 to 10RM Matrix converted into a helpful and practical tool? The 0 to 10 relationship alignment diagnostic (RAD) is the unique diagnostic tool that enables any two parties (A on figure 2.3, and B on figure 2.4 on page 52) involved in a business relationship to independently, or in collaboration, conduct a relationship health check. These two parties may be internal or external, customer and supplier, business units, work teams, functions, workplace locations or stakeholder groups, or any combination of these. The two parties could even be individuals seeking to understand and improve their working relationship.

Intuitive RAD

The intuitive RAD is an ideal relationship health check that can be conducted in meetings or workshops. It has proven to be a powerful tool to initiate a conversation and develop an improvement plan to take the journey from the current state to the desired future state for the relationship. Based on the 0 to 10RM Matrix, the RAD examines the alignment of the two relationship parties in terms of the relationship type(s) engaged and the performance level(s) achieved. It enables the understanding of the current state and desired future state for the relationship and how to bridge the gap. The RAD will also identify secondary points — the pockets of behaviour and practice that differ, for better or worse, from the current state. The intuitive RAD is completed by answering three questions:

- What is our current approach to this relationship and what performance level is being achieved?

- What is our perception of their (the other party in the relationship) current approach to this relationship and what performance level is being achieved?

- What is our desired future approach to this relationship and what performance level do we want to achieve? (Typically, but not always, the desired state is to be achieved within two to four years.)

The answers to these three questions are plotted onto the 0 to 10RM Matrix as numbers 1, 2 and 3. Answers 1 and 2 represent the current state of the relationship and answer 3 the desired future state. Secondary states, negative or positive, can also be plotted as diamonds on the 0 to 10RAD Matrix. This is known as a 1 Party RAD (also called a 1 Party RAD A) as it is completed by one of the two parties engaged in the relationship (Party A). Figure 2.3 shows an example of a 1 Party RAD.

The party completing the RAD then tries to capture the reasons the three numbers are where they are on the matrix.

Figure 2.3: example of a 1 Party RAD A

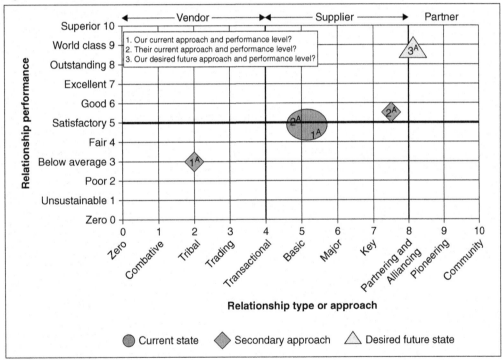

The most effective results are achieved from the intuitive RAD when both parties of the relationship complete the RAD separately and the results are analysed jointly. That is, two 1 Party RADs are completed separately by the relationship parties (known as parties A and B) and the results combined to produce a 2 Party RAD.

Figure 2.4 shows a completed 1 Party RAD for party B and figure 2.5 shows the subsequent combined 2 Party RAD (A and B) result.

Figure 2.4: example of a 1 Party RAD B

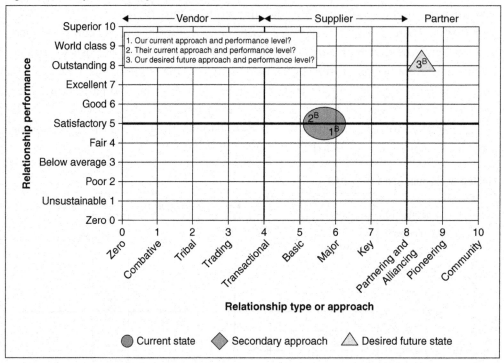

Through this collaborative two-party approach, there is a natural cross check and alignment of information, perceptions and ideas, allowing for a more comprehensive and accurate analysis. Your questions 1 and 2 are the other party's questions 2 and 1 respectively. Experience indicates that this joint approach encourages greater openness and faster, more sustainable, improvement in relationship development. In short, it is the conversations around both parties' perceptions of each other's approach that generates alignment and adds value.

The intuitive RAD is an ideal in-house diagnostic tool to reach collective internal understanding of a relationship under review. Alternatively, the RAD can be used as pre-work in preparation for a relationship review and improvement workshop and understanding alignment within the relationship parties. It can also be completed during the workshop as a group activity. Completing regular RADs (or online eRADs at <www.0to10rm.com>) for the same relationship will produce trend lines over time and can be an accurate measure in tracking relationship progress and improvement.

The RAD is just as applicable to the analysis of internal relationships between functions, departments and business units.

Figure 2.5: example of a 2 Party RAD (A and B)

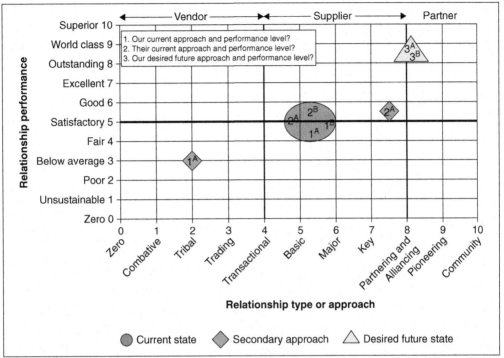

Try the intuitive RAD for yourself on the 0 to 10 RAD in appendix C. As optional pre-work to a two-party RAD exercise, answer the following five questions for the relationship under review:

- What are the *highlights* of the relationship(s) to date (the success stories, the positives)?

- What are the *lowlights* of the relationship(s) to date (the failings or problems)?

- What have been the *rub points* (points of friction or frustration in the relationship)?

- What are the *opportunities* for improvement moving forward?

- What are the *barriers* to implementation for these opportunities?

The answers to these questions, combined with the details in the characteristics matrix (see table 2.1, on pages 48 and 49), will provide useful background information for completing the RAD (that is, plotting the answers to questions 1, 2 and 3 on the 0 to 10RAD matrix).

The 0 to 10RAD result and the characteristics matrix help the parties answer three questions associated with bridging the gap between the current state to the desired future state. At the individual, team or work group and organisational level:

- What do you need to stop doing?

- What do you need to start doing?

- What do you need to continue doing?

Conclusions from the intuitive RAD results are often revealing and unexpected, generating a moment of truth or eureka experience for the participants.

e-RAD

The 0 to 10RM eRAD <www.0to10rm.com> is a unique online relationship alignment diagnostic and business intelligence tool. It is the web-based, automated version of the intuitive RAD. The eRAD is an independent, relationship health check designed to help respondents to understand both the current state and desired future state of their most important business relationships and how to bridge the gap between them. The eRAD business intelligence functionality includes:

- time trend analysis by plotting multiple eRADs

- market sector analysis

- relationship strategy maps

- external customer and supplier segmentation analysis

- internal relationship analysis

- assistance in determining the value of relationships underpinning mergers and acquisitions.

The eRAD can be accessed and completed at <www.0to10rm.com>. The advantages of employing the intuitive RAD and eRAD are threefold:

- A realistic view and understanding of both the current state and the desired future state of the relationship under review.

- From these two anchor points a current state, desired future state improvement plan is able to be developed and implemented.

- Relationship(s) progress and improvement can be monitored and tracked on a regular basis.

As a guide there is typically a 10 per cent to 50+ per cent win–win improvement to be gained in moving from the current state to the desired future state. Financial and non-financial business performance outcomes will depend on the size and nature of the improvement opportunities. More specifically, there is typically a 10+ per cent improvement in financial or non-financial metrics for every one point improvement on the 0 to 10 relationship performance or relationship type scales as measured by intuitive RAD and eRAD. Almost always the exceptions to these guidelines involve outcomes on the higher side of results expectations. The key results areas involved would include:

- financial success

- customer or stakeholder satisfaction

- sustainable competitive advantage

- best practice implementation

- innovation

- attitude.

0 to 10RM relationship strategy map

The 0 to 10RM relationship strategy map enables the mapping of your most important relationships within a single framework using the 0 to 10RM Matrix as the template. This provides a strategic overview of the current state and desired future state of the organisation's most important relationships. It is a plan on a page tool. As with the RAD and eRAD, it can be applied internally or externally to the organisation, upstream with suppliers and service providers, and downstream with customers and clients.

The 0 to 10RM relationship strategy map in figure 2.6 (overleaf) shows the current state and desired future state for nine different relationships. This strategy map is a typical snapshot of many organisations and the high impact internal and external relationships that are in place to support the delivery of specific high growth, industry segment business objectives.

Table 2.2 (overleaf) gives a simple explanation of the relationships shown on the relationship strategy map in figure 2.6.

Figure 2.6: example of a relationship strategy map

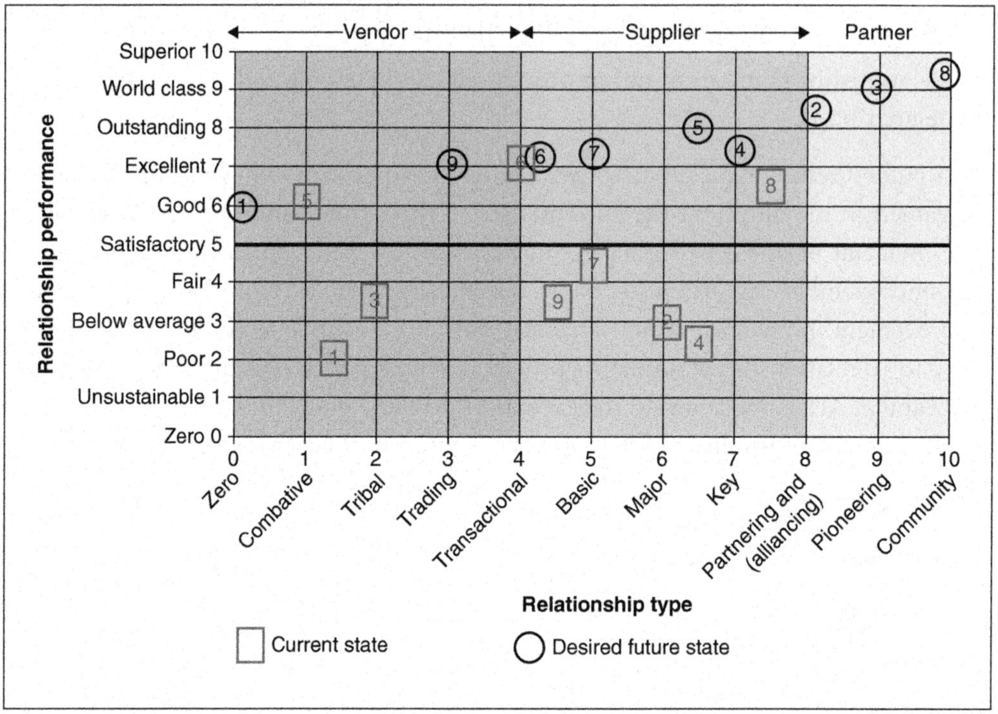

Table 2.2: explanation of the relationships shown in the relationship strategy map (see figure 2.6)

Relationship number	Comment	Action
1	Major external service provider. The current state at a poorly performing Type 1 Combative or Type 2 Tribal relationship is not sustainable. Cultures and strategies are not aligned for the longer term.	Exit strategy is required, but it must done in a way that delivers a good result for both parties. Evaluate and select new service provider(s) around strategic Type 8 Partnering and Alliancing engagement to deliver outstanding to world class performance. Ensure smooth handover and transition.
2	A critical long-term customer or client relationship—it is number one in this critical market segment and generates 30 per cent of total revenue, and this percentage is growing. The relationship is not effectively engaged and is not performing anywhere near the expectations of each party. A review and reset process is to be engaged, with strong governance and leadership required to communicate the new vision and purpose.	Below average Type 6 Major or Type 7 Key outsourcing relationship needs to move to a world class Type 8 Partnering and Alliancing relationship.

3	A critical internal operations and maintenance relationship that is currently very protective and defensive with many demarcations. Fundamental change is required if the external business strategy is to be successful. This relationship should be a role model for others (internal and external) to follow.	Move from a below average Type 2 Tribal relationship to a world class Type 9 Pioneering relationship.
4	Number one raw material supplier delivering below green line performance, with fault on both sides. The relationship needs resetting to reach a new level of support and commitment. The value propositions around research and development and new technologies support a higher level engagement.	Build a better performing, excellent to outstanding, Type 7 Key relationship.
5	Number two customer constituting 15 per cent revenue with low but sustainable growth projected. Relationship performance is above the green line but they are very difficult to deal with. This results in wasted time, resources and high stress levels. Not a sustainable position in the long term.	Review and reset the baseline for improvement and install a new relationship manager to kick start the significant shift from a good performing Type 1 Combative relationship to an outstanding Type 6 Major or Type 7 Key account within 18 months.
6	A transactional IT support relationship that is delivering excellent service and a low total cost, with long-term capability in place. This is an 'If it ain't broke we don't have to fix it' relationship.	Maintain a business as usual strategy with effective monitoring and management of service levels, response times, systems uptime and overall quality performance.
7	A Type 5 Basic *do and charge* freight and distribution supply arrangement delivering only a fair to satisfactory below green line performance.	Continue a basic approach and refocus on improving the performance of planning and scheduling processes and response times while maintaining existing cost structure.
8	Internal and external, research and development and technical support groups working well together and sharing information for mutual benefit of the broader community. There is room for improvement in both relationship approach and performance. In some cases the parties are acting as islands of cooperation rather than as a collective body that has a high social and broader community impact as well as business impact.	Set a strategy for developing a world class, Type 10 Community relationship that has a demonstrated triple bottom line effect.

(continued)

Table 2.2 *(cont'd)*: explanation of the relationships shown in the relationship strategy map (figure 2.6)

Relationship number	Comment	Action
9	A supplier of important but undifferentiated commodities used in the manufacturing process. The current Type 5 Basic *do and charge* relationship is not working well, and is delivering below average performance on price, quality and availability.	Go to market using open tender and set up a competitively benchmarked Type 3 Trading relationship to deliver excellent performance at a fair and competitive price.

The 0 to 10RM relationship strategy map, together with the 0 to 10RM RAD, or online eRAD, are effective relationship management and business intelligence tools linking individual relationships to the bigger picture strategy. The application and combinations that can be analysed are almost limitless, from internal and external, market sector, to country and regional comparisons. These tools form a core component of the broader business strategy.

The benefits of making relationship management a core competency

All too often the focus of sales and marketing is more about getting rather than keeping customers. Building and sustaining high performance business relationships will reduce customer churn and improve customer loyalty, which will open up more value-adding opportunities and reduce the total cost of ownership.

Core competencies[4] are those groups of activities, skills and technologies that a firm does well (ideally at world class performance level), which add direct value for the customer. In doing so, these competencies clearly advantage and differentiate the firm from its competitors, allowing the firm to extend itself into new markets, products and services. Understanding and developing core competencies is fundamental to any organisation striving for market leadership.

With global markets and global competition now a reality, faster access to information, new technologies, products and services are crucial in a world of dramatically reducing life cycles. High performance coordination, cooperation and collaboration are fundamental to competitive advantage and business success. In moving from vertical integration to virtual integration, customers, suppliers and partners in high performance relationships will not only exploit synergies but also leverage core competencies.

Making relationship management a core competency becomes a key strategic platform upon which superior organisational performance and competitive advantage can be developed and sustained. The 0 to 10RM models, processes and tools enable this core competency to be developed.

Summary

- The 0 to 10RM Matrix (see figure 1.4 on page 10) is the centrepiece theme image on which the other four theme images are based. It has direct application to all business relationships, internal or external, in any market category, and in both the public and private sectors.

- The 0 to 10RM Matrix provides a relationship framework for understanding:
  - the two anchor points — the current state and the desired future state for any relationship.
  - secondary states (those outliers — pockets of behaviour, practice or performance — that are different from the current state).
  - the 'how to' journey roadmap to bridge the gap between the current state and the desired future state.
  - 0 to 10RM principles 1 and 2 are directly associated with the 0 to 10RM Matrix: principle 1 where you can't be all things to all people; and principle 2 where relationship approach is firstly a choice and then a responsibility.

- The 0 to 10RM Matrix outlines the relationship choices and is composed of:
  - 11 relationship types spread across three segments (vendor, supplier, partner), plus the Type 0 Zero relationship type.
  - 11 performance levels divided by the horizontal line on the matrix (see figure 1.4) at performance level 5. All business and organisational relationships fit somewhere on the 0 to 10RM Matrix.

- The five relationship components, or characteristics, to be aligned with each relationship type are culture, strategy, structure, process and people. There are six key results areas (KRAs) to be measured. These are financial success, customer and stakeholder satisfaction, sustainable competitive advantage, best practice implementation, innovation and attitude.

- There are two steps in understanding which relationship approach is appropriate:
  Step 1: Application importance — the relationship between strategic value and commercial value.
  Step 2: Organisational trustworthiness — the relationship between willingness and capability.

- There are two practical applications or tools associated with the 0 to 10RM Matrix: the 0 to 10RM RAD (relationship alignment diagnostic) relationship health check and the 0 to 10RM relationship strategy map.

- Where you are (current state) and where you want to go (desired future state) on the 0 to 10RM Matrix will have implications for business strategy and go to market strategy; supplier evaluation and selection processes; negotiation, leadership, communication and interpersonal styles to be used; team dynamics, and people and change management; commercial and contractual frameworks, and terms and conditions; relationship manager competencies; the role of relationship facilitators; and relationship management and improvement strategies.

- Developing relationship management as a core competency will assist in delivering added value and sustainable competitive advantage.

Chapter 2 looked at and unpacked the 0 to 10RM Matrix into its components. Now for the details! This chapter examines each of the vendor, supplier and partner segments and the eleven 0 to 10RM relationship types and takes a deeper look into their specific qualities with accompanying examples and stories.

The 11 relationship types on the 0 to 10RM Matrix are illustrated in figure 3.1 (overleaf). This figure presents the circular view of Type Zero relationships and all other relationship types within the vendor, supplier and partner segments. The supporting details surrounding the three segments as well as the Type Zero relationship are discussed on the following pages. Figure 3.1 also prompts us to think about the next relationship type to be discovered.

By definition, the Type Zero relationship is non-existent, indirect or aspirational. This relationship type lives outside the vendor, supplier and partner segments. The Type Zero relationship does not have a natural grouping with the other 10 active relationship types.

A relationship outside the three segments

Type Zero relationships occur where the choice is made, deliberately and consciously, for good reason(s), not to have a relationship with the customer, supplier, stakeholder or competitor in question.

Type Zero relationships

As strange as it may seem, the Type Zero relationship is not only legitimate but also essential to your business strategy and success. First, there are those organisations that you choose deliberately and consciously not to do business with. These could be customers, suppliers, competitors or other organisations. The reasons for this choice may be ethical or commercial in nature, or there may simply be little or no strategic alignment with the organisation. The circumstances could involve a lost account where the relationship will be left at Zero (see figure 3.2), or deliberately discarded business. It may be to your advantage to walk away from these relationships and let your competitors invest their time, money and resources on difficult or unproductive relationships.

Figure 3.1: the 0 to 10RM relationship types

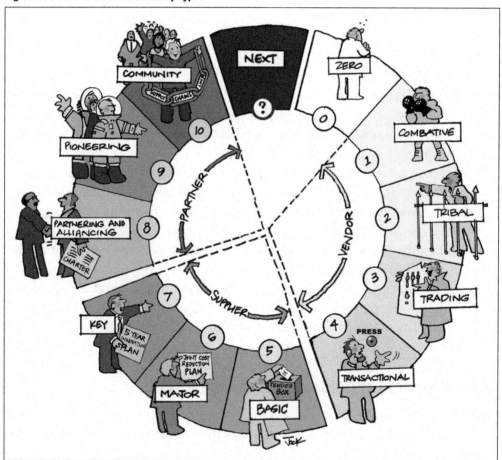

For other customers or suppliers, the Zero relationship approach could involve potential new business accounts or lost accounts to be regained. There is a complete skill set to be engaged around this aspect of Zero relationships, such as business development, network building, information gathering and sharing, strategy development and accumulating market intelligence.

Figure 3.2: Type Zero relationship

Zero relationships also apply to third-party relationships that are managed through intermediaries. These Zero relationships can still have a significant effect on your life and the wellbeing of your business, for better or worse. This relationship could include those with government departments, law makers, regulatory bodies, other agencies, vested interest groups, subcontractors and individuals. Often their interaction with you may be through a transactional medium, such as print, television and other multi-media and third parties. Although no direct association exists, these organisations and relationships can, by their action or inaction, impact your organisation.

Zero relationships could also involve implementing a strategy to remove or terminate a current state relationship that is underperforming, perhaps because their associated products and services are being made redundant or do not align with your future strategy or direction. In this event, the quality of the exit for the incumbent supplier is as important as the effective transition and engagement of the new product supplier or service provider. The exit strategy from a current state to a Zero relationship type must also ideally be above 5 on the performance scale; that is, the green line.

An oil and gas story

I am reminded of an oil and gas producer who made the strategic decision to exit a long-standing (more than 20 years) supplier of maintenance services in order to engage a longer term strategic alliance partner for all their maintenance and shutdown activities. While the incumbent was not performing badly, the organisations had a Type 5 Basic relationship approach, a *do and charge* contractual engagement around a prescriptive, input-focused scope of work. This more traditional approach, which, by the way, both parties had supported in the past, would not sustain the oil and gas producer's future competitive advantage nor could it support the producer's future business expansion and market growth aspirations. The world was rapidly changing and they needed to stay ahead of the competition curve. The new Type 8 Partnering and Alliance approach was based on the value propositions of gaining faster access to new technologies; removing complexity and reducing operating costs by 25 per cent; accelerating improvements in reliability, planning and scheduling; and implementing greater integration, simplification and efficiency of operations at reduced risk.

It was not that the outgoing supplier had done a bad job in the past but rather that they did not have the level of willingness and capability to assist in building the future. The new strategic maintenance partners would give the client organisation the competitive edge that was required.

The client organisation, through a go to market, open bid process had made a clear decision to exit the incumbent supplier relationship in the same manner they would transition and enter the new alliance partnership. There was open, honest and transparent communication and information sharing as to *where* the new company strategy was taking them; *why* this change to a new strategic partner was important and necessary; *what* needed to be done to make the transition safe, incident-free, effective and beneficial for all parties, including the outgoing maintenance service provider; *who* was involved and affected; and *when* the initiative would start and the transition period finish to then start with the new supplier partner. The result was a delighted client; a successful, seamless, incident-free transition to a committed and enthusiastic strategic partner; and an exiting incumbent who was positively recognised for its contribution and given every support and reference to continue to successfully carry on its business. Although the change was a disappointment for the outgoing supplier, the change gave the best possible result all around.

The message is that relationship exits do not have to be brutal, alienating affairs where incumbents are removed, protesting their unfair treatment with lawyers and contract specialists in tow. The result is often expensive; causes over-schedule transitions, and slow and unproductive startups; and a legacy of bitterness and desire for revenge.

In this story, all parties positively gained from the experience and the associated change process. During an exit interview the incumbent supplier expressed their thanks for the open process, and the recognition and support they had been given, during what was clearly a difficult time. They mentioned that they had genuinely learnt from being part of the process and this would put them in a better competitive position for sustaining and building business in the future. Both the client and outgoing service provider wished each other well in their future business endeavours. No bridges were burnt; lots of lessons were learned by all involved; and the broader market place would be better off and benefit from the process and the experience.

Ten tips when implementing a Type Zero relationship exit strategy

Here are 10 ideas for engaging in a successful Type Zero relationship:

1 Understand and document the value propositions underpinning the exit strategy to a Zero relationship type, then prepare, seek endorsement and implement the new strategy.

2 Engage and inform all parties involved, including internal and external stakeholders, of your intention to exit the relationship and the reasons for doing so.

3 Agree on the where to from here strategy around evaluation, selection, negotiation, transition, startup and ongoing engagement of the new relationship(s) and the associated replacement products and services.

4 Ensure there are adequate protection, warranties, indemnities, defect liabilities, service support, resources, technology and so on in place to maintain legacy support for remaining products and services after the exit.

(continued)

Ten tips when implementing a Type Zero relationship exit strategy *(cont'd)*

5 Enlist the support of sponsors and key stakeholders to assist with potential roadblocks, rub points (points of friction or frustration in the relationship) in implementing the exit strategy (internal and external).

6 Review all relevant legal and contractual rights and obligations, processes and procedures that will be affected by the exit strategy and termination of the relationship.

7 Assist the outgoing party or parties to ensure an efficient, incident-free exit regarding demobilisation (of such things as assets, inventories, services, systems and processes), and transition to and start up of the new relationship(s) and associated products and services.

8 Understand what is required to help the incoming party or parties to ensure an efficient, incident-free mobilisation (of such things as assets, product and service inventories, services, systems, processes) and transition to and start up of the new relationship.

9 Develop and implement a change management plan to positively and effectively manage, both internally and externally the transition from exit to start up of the new relationship(s), and associated products and services.

10 Retain effective and appropriate people, systems and processes to maintain continuity of products and services and business performance.

Vendor segment relationships for relationship types 1, 2, 3 and 4

There are four distinct relationship types in the vendor segment Type 1 Combative, Type 2 Tribal, Type 3 Trading and Type 4 Transactional (see figure 3.1 on page 62).

As a society we are hard wired for vendor-type behaviour. Vendors in external relationships want to buy or sell things at cheap prices, do deals and make transactions. They normally have a least cost or lowest price, non-transparent strategy and mentality; they often protect and defend their positions, trading on thin margins (and sometimes larger margins); and there is often little to differentiate them in a positive way from their competition. Price or cost is the driving force behind the relationship, and customer or supplier loyalty is given a lower priority.

Traditional interfaces in vendor relationships typically set up competitive tensions around a buy and sell focus, albeit alongside branding and loyalty strategies. Vendor relationships are more about coordinating activities than genuine cooperation or collaboration. The vendor focus is weighted more on risk transfer from customer to supplier than on risk management. In fact, one of the qualities of a good vendor is the ability to change quickly from one customer to another, or from one supplier to another, in the event of non-availability, lack of dependability or pricing problems. In the case of a trader, it could be that the principal company changes agents, or goes direct to the customer, or the supply source dries up. Vendor relationships are occupied and managed by hard-nosed contract managers, sales and service representatives, traders, deal makers and the like.

Internal vendor relationships engage similar practices, attitudes and behaviours, but with a different focus. This might present as Type 1 Combative relationships between vertically integrated business units that have competitive offshoots; defensive and protective behaviours and practices around organisational silos; workarounds and deals done between operating units and support functions, such as operations and maintenance, engineering and procurement; and arms length, transactional engagement between primary activities and support services such as IT and manufacturing, sales and accounts receivable.

Vendor relationships appear in all market segments, both online and offline. Before and throughout the global financial crisis the behaviour of certain Wall Street banking and investment institutions, and their specific banking and lending practices demonstrated the worst of vendor behaviour—predatory win–lose, high risk behaviour; defending positions and protecting incentive schemes; doing bad deals based on self-interest with high transactional costs and low customer focus. On the other hand, very successful commodity-based businesses, and retail and online trading and transactional businesses operate in the vendor space professionally and ethically, and are both enduring and successful.

Good vendors know their business. They understand their suppliers and customers, and their requirements, their products, services and costs. These vendors research their competitors and understand points of difference. High performing vendors can have very successful business relationships, albeit with a correspondingly high level of competitive tension between the parties. They have a low cost base and don't pretend to be anything more than they are. Depending on the strategy, the time and place, and the nature of the product or service, vendors can be very competitive and effective. A good vendor may cause many a reasonable supplier or a poor quality partner to come unstuck. Good vendors, like good suppliers and partners, don't over promise and under deliver, and they can be flexible and responsive in the right circumstances, when it suits them to be so. They understand their limitations and match their capabilities to customer expectations.

Type 1 Combative relationships

Combative relationships employ a winner takes all, whatever it takes, approach to managing relationships. Uncooperative, confrontational, adversarial, divisive, aggressive and, on occasion, coercive, belligerent and pugnacious are some of the words and behaviours that sum up Type 1 Combative relationships (see figure 3.3). They are driven by a win–lose, master–slave control and compliance mentality, manipulation of facts, the need for secrecy and a short-term profit focus. This is a world of survival as much as it is about self-interest at the expense of others. Combative relationships are associated with an abrasive, taking culture, and hierarchical and bureaucratic structures that produce hostile interfaces and abrasive personal rub points. The combatants in these relationships are reminiscent of schoolyard bullies. Communications are often threatening and aggressive, creating an environment of intense competition and rivalry, internal and external, that often stifles creativity, openness, transparency and honesty. It is said that truth is the first casualty of war; Combative relationships in business are no different.

These are the mean, deceitful, vindictive, vengeful bad boys of the 0 to 10RM scale. Legal fees are one of their largest expenses, and the legal system and traditional contract process their playing field. Relationship management is seen as a blood sport. Tenders and the competitive bid process are commonplace, and are often associated with detailed, hard-nosed, hard-dollar, tightly managed, one-sided contracts. There is a strong focus on punitive contractual and behavioural drivers, such as liquidated damages, security guarantees, imposition of unlimited liabilities, performance penalties, consequential loss or similar imposts.

Combative relationships exist everywhere. We are surrounded by them. Daily, online and offline, the media salivate at the opportunity to tell us about the latest corporate melee or political confrontation. Our legal system thrives on them. Combative relationships can include hostile takeovers, mergers, acquisitions and joint ventures; many management–union relationships; competitors forced to do business with each other through legislative requirements or market forces (for example, telecommunications); hard-nosed contractual relationships, such as in the building and construction industry; rival company directors, fighting it out in the courts for boardroom control; politics and politicians using cat and mouse tactics; relationships between feuding neighbours; the divorce courts; interdepartmental rivalry; and a multitude of other business and interpersonal relationships. They range from the very large to the very small. Unfortunately, many large capital projects—for example, in the civil engineering and construction sectors—are still based on combative relationships. I knew a CEO who regarded his office as a trophy room, where he could show off on its walls the verdicts and associated publicity from successful litigation and courtroom battles.

Figure 3.3: Type 1 Combative relationships

I am not suggesting that Combative relationships can't be either appropriate or successful, but they do demonstrate less sustainability. Many successful monopolies, semi-monopolies and bigger institutions, especially in banking, information technology (IT), manufacturing, construction and telecommunications, still fly a combative flag. Risk is often transferred 100 per cent from customer to supplier or from one party to the other, or at least an attempt is made to do so. Some banking practices leading up to the global financial crisis of October 2008 were a cogent example of typical risk transfer behaviour—for example, the securitisation of sub-prime mortgages, which transferred the risk to individual investors and other investment firms.

It is impossible to transfer 100 per cent of the risk. First, even if it were possible, you would pay for the privilege. Second, there will always be some aspect that comes back to bite one or both parties, if only by affecting reputation, credibility and future competitive advantage.

A typical strategy or tactic in a Combative relationship would involve one or more of the parties employing lawyers to scrutinise contracts for potential loopholes, before, during or after they are signed or awarded. These loopholes are then used to introduce variations or scope creep changes at a point in the contract or process timeline where it would be difficult, if not impossible, for the customer or principal to displace the contractor or vice versa, or at least renegotiate the outcomes.

Being Type 1 on the 0 to 10RM scale, combatants are the partners from hell. Generally, relationships improve the further to the right you move on the scale. By default, then, this is as bad as it gets, certainly in terms of loyalty, collaboration, openness and trust. Often Type 1 Combative relationships can be a catchall for relationships that have gone wrong. Even with high performing Combative relationships, one or more parties in the relationship will make a comment like, 'They do a good job but, damn, they are difficult to do business with. They are expensive, arrogant and adversarial. They always seem to want an argument. It's very difficult to trust them fully'.

Typical negotiation techniques include power plays, good guy–bad guy plays; threatening, coercive, intimidating behaviours; bullying and heavy-handedness; requiring everything in writing; cat and mouse deception and delaying tactics, and aggressive negotiation over the contract details; hostile and coercive communication and deception; the use of misinformation, as well as negative advertising and public, and private ridicule of the other party or parties.

So why would anyone want to engage in a Combative relationship or behaviour, and why do these adversarial relationships survive for so long in many cases? There are two primary reasons.

First, there may be few alternatives, since the changeover and transition costs are significant. In many cases one alternative is as bad as the other. Turning the other cheek is not an option. If one party in the relationship is determined to be combative, then on occasions there will be few alternatives but to match like with like. Telecommunications, banking and the resource sector are typical examples.

The second reason is that Combative relationships are often cyclical relationships that ride the rollercoaster of supply and demand imbalances, and boom and bust cycles — such as the resources, coal, oil and gas, real estate and commodity sectors. These cycles often reignite the get mad and get even mentality developed in a past engagement, and adversarial practices. I remember a business manager on more than one occasion saying at the bottom of the cycle, 'I won't forget what they are putting us through. The market will turn and when it does the shoe will be on the other foot, with a significant multiplier effect.'

Some individuals are brutally competitive by nature. They just love a good fight. Old habits die hard and they thrive on competitive tension, both internal and external. Second place is seen as the first place for losers!

The Combative environment is a tough and brutal space to work in. If used well for the right reasons, these relationships can be very successful. Be careful, however; fatigue and burnout rates at a personal level are high and the retention of high performing, loyal and committed employees is problematic and often connected to misaligned incentives. Winners are grinners, and losers are part of the spoils of battle, quickly consumed and relegated to the dust bin of history.

Ten tips for implementing and sustaining a Type 1 Combative relationship

Here are 10 ideas for engaging in a successful Type 1 Combative relationship:

1 Negotiate, and enforce to your benefit, all of the techniques of the Type 1 Combative relationship such as being hard-nosed and unfriendly; managing contracts ruthlessly and acting only to the letter of the contract rather than in the spirit or intent behind the contract.

2 Communicate, verbally and in writing, in a style that is intimidating, threatening, abrasive, hostile and uncooperative.

3 Engage practices, processes, procedures, mindsets and behaviours that are secretive, deceptive and non-transparent as appropriate, for instance by creating confusion and frustration with time-wasting and costly paper trails.

4 Create an environment, internally and externally, that encourages competitive tension.

5 Run tough, hard-nosed, confrontational negotiations where concessions are ruthlessly fought over and traded and where win–lose outcomes may be seen as more beneficial than a win–win result. Use combative negotiating techniques, such as ambit claims, cat and mouse tactics, time delays, and creating deadlocks and stand-offs. Alternatively, read Sun Tzu's *The art of war* for guidance!

(continued)

Ten tips for implementing and sustaining a Type 1 Combative relationship *(cont'd)*

6 Identify, engage and train people who are comfortable working in combative relationships.

7 Establish bonus, commission or other incentive schemes that reward and recognise people for combative behaviour and practice.

8 Establish clear lines of authority and empowerment levels to maintain control and compliance.

9 Document what success looks like in moving to a Combative approach, for instance in terms of vision and mission, financial success, customer and stakeholder satisfaction, sustainable competitive advantage, best practice delivery, innovation and attitude.

10 Agree on the value propositions that underpin the Combative relationship. These are often based on dominance, such as market dominance, market share, brand dominance, power and influence, and are financially focused.

Type 2 Tribal relationships

Type 2 Tribal relationships (see figure 3.4) are often spoken about as fiefdoms or organisational silos within which exist parochial, insular, protective, defensive territorial groups that are resistant to change and have an intense suspicion or fear of outsiders. They are typically inward-focused, risk-averse and secretive relationships, and they can occur between departments, functions and other operating groups in an organisation, as well as between customers, suppliers and other organisations. Tribal relationships also extend to political, union, sporting, suburban and family environments. For example, politics is a haven for vested interest groups and factions driven by legendary turf wars and intergroup rivalry. As the old political saying goes, 'in the race of life, always bet on self-interest'.

Those involved in Tribal relationships exhibit a willingness to protect and defend everything, from territory, boundaries, departments, markets, position and title, information and power, to profits and margins. They are, therefore, often laden with hidden agendas. Within the normally hierarchical structures where these relationships thrive, there is much back watching, self-interest and finger pointing, with medium to high levels of loyalty to the group or tribe in question, but little trust or loyalty towards the external environment. Old boy networks are commonplace. There is a strong desire for ownership, of intellectual property, assets and resources, and self-interest based what's in it for me (WIIFM) attitudes and behaviours often prevail.

Figure 3.4: Type 2 Tribal relationships

But even in this environment, Tribal relationships can surpass the median, satisfactory performance level and beyond. For example, state and federal governments from different political parties are forced to cooperate on specific issues, often delivering successful and even breakthrough negotiated outcomes. Health care is an example. Many sporting, Indigenous, environmental and vested interest groups are very successful in protecting and defending their culture, heritage and environment for all the right reasons. Tribal relationships are often laden with demarcations and boundary issues where control and influence is tightly managed. Managers and leaders may be referred to as warlords in terms of the power they have over the people, business and activities they control.

Certainly, if provoked, participants in tribal relationships can quickly turn combative, but their natural instinct is to take a defensive rather than offensive approach to the external environment.

Negotiations are often conducted in series or parallel to protect information. Negotiation steps are compartmentalised so that one team, department or function has restricted access to information. The secret of the KFC or Coke recipes or the Microsoft operating system computer code is in the hands of only a few. The information is compartmentalised to ensure security, secrecy, confidentiality and integrity of information.

I am reminded of a very large and well-known consultancy firm that had an ongoing relationship with a government defence department. The work was high value, producing high impact, sensitive data and intelligence, where confidentiality and security of information were critical and could not be compromised. The relationship was based on three significant contractual conditions. The first condition was that ownership of any information and intellectual property produced during the course of the assignment immediately became the property of the defence department. The second condition was that all information associated with the assignment was to be physically or electronically destroyed or returned to defence on completion of the assignment. In some cases this involved the supervision and observation by defence personnel. The third condition was that the good, ongoing working relationship that had developed between the two organisations producing excellent results for the client was not to be used as a point of reference, referral or advertising by the consultancy firm.

By their very nature some government departments, such as social security, health, child protection and support, witness protection programs and, in particular, defence, are in the business of defending and protecting people, assets and information. In the private sector, remote mining, and oil and gas sites are examples of workplaces where Tribal relationships can flourish, for better or worse. Their history is often one of confronting adversity, and working in tough conditions where individuality, independence and heroism are recognised and rewarded. The hero mentality becomes a core cultural component of the Tribal relationship. For the people involved, high performing Tribal relationships are either a fact of life or even an end goal. This can be problematic, for example, when fundamental change is required to a meet global reality, when growth aspirations and the need for a more cooperative and collaborative culture built around integrated, trust-based partnerships and alliances, internal and external to the group or organisation, are needed. Often the challenge lies in the task of managing the difficult balance between trust and transparency and the differing expectations and objectives of stakeholders.

As a general rule, if all the parties involved know the rules of engagement up front and Principle 4 (the right people doing the right things, the right way, at the right time for the right reasons will deliver the best possible outcomes) is actively in place, then

Tribal relationships can be successful. However, even Tribal relationships performing at levels beyond the green line need to be flexible to adapt to changing circumstances and the potential to engage other relationship approaches in achieving better outcomes.

Type 1 Combative and Type 2 Tribal relationships are often the subject of secondary points on the 0 to 10RM Matrix — that is, they are not the prime relationship approach or the current state, but rather outlying attitudes, behaviours, practices or performance outcomes are Combative or Tribal in nature. These may have a negative impact on the main relationship, especially if the desired future state is further to the right on the relationship type scale.

Ten tips for implementing and sustaining a Type 2 Tribal relationship

Here are 10 ideas for engaging in a successful Type 2 Tribal relationship:

1   Agree on clear lines around job or role responsibilities and accountabilities within the group in order to protect or defend your position or communicate demarcation lines.

2   Identify key influencers within and outside the group, their roles and how they can help in achieving the group's goals and objectives. Recruit people who are fiercely loyal to the group and will defend and protect the group interests.

3   As appropriate, restrict, manage and control internal and external communication and information sharing to maintain or enhance group position.

4   Review contractual obligations or implications in moving to a Tribal relationship. For example, the transfer or management of risk, information sharing and communications, contractual or legal exposure.

5   Identify the outsiders and protagonists who are misaligned with your position, and implement actions to positively influence or manage them.

6   Build loyalty and trust within the group through authoritative and protective leadership styles and the enforcement of clear group norms and rules. Instigate financial bonus schemes, positive reward and recognition programs for above-target performance, as well as acts of loyalty and support for the group.

(continued)

Ten tips for implementing and sustaining a Type 2 Tribal relationship *(cont'd)*

7 Review legal obligations in moving to a Tribal relationship approach to ensure protection of the organisation's rights and entitlements.

8 Be more protective of the group's information, both financial and non-financial, including knowledge and intellectual property.

9 Agree and document the traditions, rituals, cultural traits and norms, policies, procedures and practices that you want to defend and protect. Protect and, if necessary, defend your interests, such as intellectual property, information, assets, products, services, people, departments and business units.

10 Agree on the value propositions that underpin the Tribal relationship approach as the desired future state. These propositions are often based on defend and protect positions around ownership and self-interest, information, market share, financial matters, intellectual property, power and influence.

Type 3 Trading relationships

Type 3 Trading relationships (see figure 3.5) exist in a world of short-term opportunism and are all about negotiating, bargaining, bartering, contra arrangements, horse trading, deal making and getting or giving the order, predominantly at the best or lowest price. Quality and service aspects are normally to minimum standards and often a lower priority. This is the classic environment of low margins, low loyalty and little differentiation of products and services, and a focus on performance against budgets and other short-term profit and financial targets. Trading relationships tend to be shallow, short-term relationships where the speed and timing of decision making is critical. One-time only or irregular buying patterns are just as common as trading ongoing, repeat business. Trading relationships frequently involve formal and informal workarounds or networks where the emphasis is on negotiating or doing the deal, retaining existing business and getting or giving the purchase order at the best or lowest price.

Intensely competitive, Trading relationships are found everywhere from the share trading and currency markets, and traditional commodity markets, such as mining, oil and petroleum products, agriculture, petrochemicals, industrial chemicals and energy markets, through to traditional selling organisations, such as car and property sales, and the food, homewares and clothing markets. However, there is always opportunity to make a difference in any of these market places. In the words of Theodore Levitt, 'every product and service is differentiable'.[1]

Figure 3.5: Type 3 Trading relationships

Trading relationships are often associated with best-of-three-quotes tender processes and competitive bidding, in which one organisation is traded off against another for the best deal. Sales representatives, purchasing officers, sales, marketing or procurement specialists, contract specialists, traders, deal makers and the like are typically the main or single points of contact.

One example of a Trading relationship comes from the petrochemical sector, where ships loaded with ethylene, the base chemical for plastic manufacturing, are often used as ocean-going storage vessels. While many vessels have a destination booked before they leave port, others set sail in the expectation that a deal will be done somewhere on the journey. The supply and demand balance is such that ships are often redirected to a new destination based on a new or better deal being struck. Commodity-based food organisations, chemical manufacturers, oil and gas companies and energy producers employ highly paid, skilled specialists at their trading desks, which are staffed 24 hours a day. Trading desks have a global reach, buying and selling commodity-based products and they can be very successful businesses.

Trading relationships can exist internally between different departments and functions in an organisation. Operations and maintenance divisions typically involve a world of workarounds, where deals and trading concessions are negotiated around such things as response times, plant and equipment uptime and downtime, routine maintenance requirements, the scope and timing of shutdowns and turnarounds, and the timing of minor and major projects. Marketing, sales and procurement functions are also commonly involved in these arrangements to align the right supply and demand balance between the availability of input, such as raw materials, resources, plant and equipment, and the required production output. The result is a deal that everyone can live with, based on traded concessions and compromises.

Ten tips for implementing and sustaining a Type 3 Trading relationship

Here are 10 ideas for engaging in a successful Type 3 Trading relationship:

1 Build trust and credibility by honouring promises and commitments. Meet or exceed agreed product and service requirements In Full On Time to A1 specification (IFOTA1), irrespective of whether you are the (internal or external) customer or supplier in the relationship.

2 Establish single-point or face-to-face contact, responsibility and accountability within business units, functions and departments, supported by appropriate electronic interfaces.

3 Ensure contracts or internal service level agreements (SLAs) are tight and prescriptive in the specification of quality, performance and delivery.

4 Focus on and identify the product or service features that can be linked to benefits. Deal with customer or supplier scepticism, misunderstandings, deficiencies and shortcomings.

5 Prioritise the core non-negotiable issues and the concessions or features to trade to your advantage for informed, mutually beneficial outcomes.

6 Review the people, competencies and skill sets required to engage in a Trading relationship.

7 Be opportunistic: proactively seek short- and long-term opportunities to take advantage of quick wins and longer term deals. Use traditional negotiating techniques to extract the best possible deal, such as ambit claims, delaying tactics, parallel negotiating, barter, trade, contra arrangements, volume discounts, extended term incentives.

8 Seek out opportunities to differentiate your products and services. Counter a low-loyalty, undifferentiated environment associated with commodities and other traded products and services using such techniques as bundling features into benefits, aggregate purchasing and value adding to sweeten the deal.

9 Employ, develop and seek out excellent, outstanding, world class deal makers to win and grow the business.

10 Document what success looks like in moving to a Trading approach, for instance, in terms of financial success, customer and stakeholder satisfaction, sustainable competitive advantage, best practice delivery, innovation and attitude.

Type 4 Transactional relationships

Traditionally, Type 4 Transactional relationships (see figure 3.6, overleaf) involve the purchase or sale of products, services and information over the counter, over the phone, by fax or over the internet, where little or no negotiation is involved. The Trading relationship is characterised by a take it or leave it mentality, built around standard terms and conditions, as well as low cost, high speed and global reach of transaction. The characteristics of Transactional relationships are quite different from those of Trading relationships. People may or may not be directly involved in transactions, which will focus on the sale of goods or services for a set price, with point of sale payment or a predetermined set of payment terms and conditions. Although Transactional relationships employ service-oriented staff, no lengthy communication is entered into, nor are personal relationships built. Dealing with supermarkets, retail shopping outlets and many government agencies are familiar examples. Transactional relationships can be found in every aspect of both the public and private business sectors.

Transactional relationships have become far more prominent over the last 10 years, led by the exponential growth in electronic, mobile and internet communications.

Electronic funds transfer, supply chain management systems, internet marketing, online transaction processing, business process outsourcing, electronic data interchange, management information systems, inventory management systems, automated data collection systems, business intelligence applications, online track and trace systems are all part of modern electronic commerce that are conducive to Transactional relationships. In many cases, relationships have become impersonal and arms length, driven by systems or technology. These relationships are often faceless and invisible.

Figure 3.6: Type 4 Transactional relationships

E-commerce and working through sophisticated and often outsourced call centres, internet-based banking activities are typical examples of modern-day Transactional relationships. This creates the perception, if not the reality, of a cold, clinical, technology-driven and impersonal approach to relationships. Transactional relationships can, however, be very effective, efficient and profitable.

Transactional technologies have evolved and not only give valuable support to all other 0 to 10RM relationship types but also allow the evolution of a more personal, interactive and high performing transactional engagement. Not only are Transactional relationships legitimate relationship types in their own right, they are also key support elements of a broader business strategy. For example, internet banking, online booking for the airline and accommodation sectors, intranet and extranets supporting supply or value chain management, online training, e-procurement, customer relationship management systems and other business intelligence applications can play key support or secondary roles in other relationship approaches, ranging from Type 1 Combative to Type 10 Community.

Ten tips for implementing and sustaining a Type 4 Transactional relationship

Here are 10 ideas for engaging in a successful Type 4 Transactional relationship:

1 Review and optimise fees, charges and cost structures based on a Transactional relationship.

2 Build trust and credibility by honouring promises and commitments. Meet or exceed product and service requirements In Full On Time to A1 specification (IFOTA1), irrespective of being the (internal or external) customer or supplier in the relationship.

3 Apply standard terms and conditions to customer and supplier agreements, thereby avoiding the need to negotiate price, terms and conditions, and products and services.

4 Use systems and technology to increase the speed and efficiency of transactions (online and offline), reduce total transaction costs, eliminate or minimise the need for personal contact, and improve fault finding, and the quality and flexibility of product or service delivery.

5 As appropriate, minimise the face-to-face component and other non-Transactional relationship characteristics of your product or service. For example, introduce a greater level of internet and electronic interaction; and establish a help desk or call centre facility to handle customer and supplier contact.

6 As appropriate, develop online, arms length, bid systems based on impersonal tender or procurement, and processes for agreed market segments, such as reverse auctions.

7 Establish a service capability for information access, payments, problem solving and troubleshooting online or by telephone. For instance, use remote telemetry and other technologies to

(continued)

Ten tips for implementing and sustaining a Type 4 Transactional relationship *(cont'd)*

automatically manage the ordering and delivery processes, inventory management systems and condition monitoring applications to drive down the cost of transactions, improve response times and drive up value-add and customer and stakeholder satisfaction.

8 Benchmark for transactional capability against industry norms or world class benchmarks in areas such as cost, price, service levels, systems uptime, speed of transaction, and business intelligence capability.

9 Review the people, competencies and skill sets required in a Transactional relationship. Seek task-driven, service-oriented people who are comfortable with and competent in the use of technology. Employ and train staff in high level Transactional relationship techniques, such as online and offline buying and selling.

10 Agree on the value propositions that underpin the Transactional relationship, which are typically based on speed, efficiency, low transaction cost, standard terms and conditions, systems, processes and technology benefits.

Supplier segment relationships for relationship types 5, 6 and 7

Supplier segment relationships involve the cooperative, supply and procurement of products and services around progressive *do and charge* (Type 5 Basic), *do and improve* (Type 6 Major), and *do and add value* (Type 7 Key) strategies, all of which are based on common interest. Delivering products and services IFOTA1 specification becomes the base requirement for any reasonable type 5, 6 or 7 supplier. This behaviour is based on common interest, not necessarily mutual benefit. Understanding the customer's requirements, meeting or exceeding them and servicing them through cost-reduction and value-adding initiatives, such as technical service, product development, maximised synergies and responsiveness to special requests, become a good supplier's trademark—all at a competitive price. Usually, but not always, these relationships externally involve tightly managed contracts, often detailed and associated with tenders and competitive bidding. The degree of innovation in these initiatives will determine the degree of differentiation achieved, and, in turn, profitability. When relationships are moving from 5 to 6 to 7 on the relationship scale, there tends to be a growth in the size, complexity or importance of the relationship.

Many organisations believe themselves to be good suppliers and good customers positioned at types 6 and 7 on the relationship scale but, in fact, they are nothing more than vendors with type 1, 2, 3 or 4 relationships. Delusions of grandeur can lead to a fall into the well of mediocrity.

Most businesses today are striving to become good suppliers and to have good suppliers working for them. High performing type 5, 6 or 7 relationships represent very effective, competent, quality-driven and probably continuously improving organisations. Many outsourcing relationships are part of the supplier segment. For many customers and suppliers today, high performance relationships in the supplier segment will sustain a sound and profitable business and strong customer and supplier relationships.

Using the maintenance analogy, Type 5 Basic relationships involve mainly reactive breakdown maintenance. Type 6 Major relationships would involve reducing overall maintenance costs against an agreed baseline. Type 7 Key relationships would involve proactively improving overall reliability and availability of plant and equipment.

Supplier segment relationships have a customer focused, continuous improvement culture; cost or value differentiation strategies; ways of managing the push–pull tension associated with increasingly complex multi-level interfaces; processes focused on supply and procurement; and management by Basic, Major and Key account managers and project managers (see figure 2.1 on page 37).

Type 5 Basic relationships

Type 5 Basic relationships (see figure 3.7, overleaf) *do and charge*, based on an agreed scope of work, with a prime focus on competitive price, quality and measured delivery of agreed products and services IFOTA1 specification. Basic relationships are typically reactive in nature and input-driven. The Type 5 Basic relationship is the first of the genuinely customer focused or serviced or account-managed relationships beyond the spot buying and selling, and regular dealing of Type 3 Trading relationships, and the often impersonal, faceless detachment of Type 4 Transactional relationships. Basic relationships normally involve simple, single or limited points of personal contact.

The environments in which Type 5 Basic relationships exist are highly competitive. They are often initiated by referral from another organisation on the basis of the company's previous good performance, or by winning a tender and competitive bidding process. These relationships are managed through company, industry or professional standards, and associated agreements, or by inflexible, prescriptive specifications, contracts or SLAs.

The focus is on the short- to medium-term, and cost performance against budgets. Type 5 Basic relationships are independent, often reactive, work to rule and task-driven; there is little focus on or requirement for innovation or continuous improvement other than through changing with general market trends.

Figure 3.7: Type 5 Basic relationships

Electrical, mechanical, building and civil trades and many professional services, such as legal, medical, dental, engineering, scientific, financial and accounting, are typical examples of Type 5 Basic relationships. Small accounts and small- to medium-sized projects; service organisations such as office and stationary supplies, computer support firms and general consultants are also likely to operate in Basic relationships.

There is no shame in being involved in a high performance Basic relationship. But Basic relationships aren't about sharing detailed strategy, sharing risk or reward, or being at the front of technology development or innovation, even though these initiatives may be in place with other relationships the organisations might have. The focus in a Type 5 Basic relationship is on getting the job done, delivering a basic product or service that, when done well, delivers a fair return on investment and value for money for all parties involved.

These relationships, done well for the right reasons, can be sustainable and successful. They thrive on referral and repeat business, and often constitute a large segment of a typical organisation's relationships. They will normally operate in association with small to medium-sized accounts where the products and services are not critical in application, or large in size or dollar value to the customer.

Although Type 1 Combative and Type 2 Tribal relationships may share some of the same contract and work scope characteristics, they are quite different from Basic relationships. Basic relationships are not characterised by the same aggressive, adversarial, confrontational, threatening or hostile behaviours that typify Combative relationships. Nor do they display the degree of territorial, protective or defensive attitudes or approaches that are so much a part of Tribal relationships.

Type 5 Basic relationships are relatively simple relationships of low to mid-range importance, and short-term to medium-term focus, managing repeat business for small to medium-sized accounts and projects. They are serviced by task-driven basic account managers, project managers, customer service managers or people in similar roles who deploy reactive relationship management skills. A simple action plan is likely to be in place with basic key performance indicators (KPIs) to measure performance.

Basic relationships are often referred to as list price or small discount off list price, standard rates, no frills relationships that offer proactive customer service and reactive product or service development. All too often, however, Basic relationships result in medium-sized to large customers and suppliers feeling ignored, neglected, poorly treated, under-utilised, disenfranchised and generally unloved. Basic relationships are likely to characterise organisations whose products and services, skills, expertise and value-adding potential could deliver far more than a basic *do and charge* outcome. These organisations have so much more to give, but they are not given or don't take up the opportunity.

We can all relate to Type 5 Basic relationships in our personal and business associations. These relationships thrive on clear requirements (such as product or service specifications) and performance expectations (such as IFOTA1 specification service levels). If these requirements and performance levels are met or exceeded, there is the correct expectation that the invoice will be accurate and in turn paid In Full and On Time. Satisfied customers are then happy to refer the supplier organisation to others. There is no expectation of the organisation being treated as a strategic partner who has access to leading edge or bleeding edge innovation and technology, and the two organisations in the relationship have no interest in sharing each other's broader strategic goals and objectives. That does not mean that all parties cannot be thoroughly satisfied from the experience of working with each other.

Ten tips for implementing and sustaining a Type 5 Basic relationship

Here are 10 ideas for engaging in a successful Type 5 Basic relationship:

1 Focus on competitive price, basic quality specifications and measured delivery of agreed requirements (In Full On Time to A1 specification) for agreed products and services.

2 Establish a relationship that is customer friendly, efficiency based and businesslike around a strategy of *do and charge* on an agreed scope of work that delivers potential return or referred business.

3 Meet or exceed delivery of products and services to In Full On Time to A1 specification to maximise potential for return business or referred business opportunities.

4 Develop a Basic action plan to deliver or receive the products and services. Track progress. This may involve an integrated plan across a segment or sector of Basic relationships.

5 Build trust and credibility by doing what you say you will do.

6 Focus on basic product or service delivery, with short-term to medium-term objectives and milestones, rather than focusing on high-end innovations or total cost improvements.

7 Consider mostly fixed price, fee for service, schedule of rates, *do and charge*–based commercial arrangements rather than more sophisticated gainsharing and painsharing commercial frameworks that are constructed on formalised risk and reward sharing.

8 Have in place clear, tightly managed, plain English contracts or SLAs that are prescriptive and simple to effectively manage the relationship. Include a clear set of predominantly input-focused product or service specifications in contracts.

9 Agree on the performance levels required and have in place a simple, clear set of compliance or basic KPIs, based for instance on quality, cost, price, schedule, and In Full On Time to A1 specification delivery.

10 Define Basic account manager or procurement roles. Document roles and responsibilities, and authority or empowerment levels for managing relationships at a Basic level.

Type 6 Major relationships

Type 6 Major relationships (see figure 3.8) are typified by an increasing complexity and significance of the relationship and the associated products, services and projects delivered. Type 6 Major relationships *do and improve* against agreed baselines. They focus on total cost reductions, over and above the IFOTA1 specification delivery of agreed requirements. They are the first of the authentic, proactively managed, results-driven, customer-focused, total quality-based, continuous improvement relationship types.

Figure 3.8: Type 6 Major relationships

Major relationships are characterised by a sharing of business goals, objectives, performance drivers and measures, relevant financial and non-financial information, and opportunities for improvement. There is a medium-term to long-term focus oriented to reducing total costs over and above adding value.

Major relationships are usually independent rather than interdependent relationships, and they are characterised by minimum sharing of risk. Within this relationship type, business goals, and performance drivers and measures are shared. Individual corporate objectives are identified and reviewed to ascertain where they complement or conflict with each other. The relationship parties typically conduct a joint SWOT (strengths, weaknesses, opportunities, threats) analysis within a medium- to longer term perspective.

A formal differentiation strategy involving process re-engineering or new product or service development is in place, based on total cost reductions against agreed baselines. Adding value in terms of new product and service development to improve the customer's margin, selling price, sales volume or market share tends to be more reactive than proactive. Results-driven major account managers, project managers, asset managers, procurement supply and contract managers actively manage and monitor the progress of these relationships. Monitoring includes regular reviews of the overall supplier performance. Ideally an informal, multi-level network of internal and external service providers would support the relationship managers. Major account plans are in place, and these are linked to KPI measures and contractual obligations, and supported by a medium level of multi-contact and contract management interfaces.

Many outsourced relationships of non-core activities would be candidates for Type 6 Major relationships. In fact, Type 6 Major relationships are often seen as the entry point for outsourcing — outsourcing being defined as the transfer of non-core competencies or activities from customer to supplier. In effect, sub-contracting the products or services to a third party. Many would involve single or preferred supplier arrangements with medium levels of systems or process integration. On occasions, Type 6 Major and Type 7 Key relationships are seen as part of best of breed strategies, where the best features, competencies and qualities are cherry picked from suppliers. At a strategic level a best of breed strategy sometimes presents itself as multiple, independent relationships delivering varying value, based on customised or proprietary products, services and technologies. Telecommunications, banking, petrochemicals, oil and gas, manufacturing and other sectors often have best of breed legacies due to past relationship and procurement strategies.

The classic phrase often used around outsourcing Type 6 Major relationships is 'We could do it ourselves but we don't have the time and want to save on costs. If we outsource we will be able to get more for less! They are the experts in that area so let them save our time and our money'. Products and services that are often outsourced under term contracts are likely to be a mix of input- and output-driven specification requirements, and financial- and service-driven performance indicators. Major relationship contracts are often associated with competitive bidding and tendering. Some examples of Type 6 Major relationships would include facilities management, catering and cleaning services, maintenance and engineering services, operations outsourcing, manufacturing, grounds management, IT support, customer support and call centres, accounting and financial services, travel and accommodation services, automotive fleet management, market research, freight and distribution, human resources, training and development, manufacturing design, and web design, development and support. Also included could be internal relationships between functions and departments, which are regularly seen as the relationship between the primary activities in an organisation's value chain (for example, operations, logistics, marketing, sales and service) and support activities (for example, IT, human resources, procurement). All market segments, small, medium and large businesses, and public and private sector organisations are potentially affected or involved. Typical KPIs involved in these relationships include:

- lowest total cost of ownership as benchmarked

- quality, schedule, service level improvement opposite baselines

- x per cent increase in new business revenues

- 100+ per cent return on investment (ROI) or other financial targets

- x per cent faster workflows or productivity and response time improvement

- 100+ per cent adherence to internal or external client SLAs.

Ten tips for implementing and sustaining a Type 6 Major relationship

Here are 10 ideas for engaging in a successful Type 6 Major relationship:

1 Adopt a KPI focus on short-term to medium-term cost, quality, schedule, service levels and performance improvements from established baselines.

2 Break down and analyse the relationship supply chain and internal value chain (primary and support activities). From this analysis agree on financial and non-financial baselines for total cost improvement targets, process re-engineering, value engineering, and people and change management opportunities.

3 Have in place a clear, plain English contract or SLAs that include product and service specifications that are both output-focused and results-driven.

(continued)

Ten tips for implementing and sustaining a Type 6 Major relationship *(cont'd)*

4 Focus on total cost improvement, quality improvement, and continuous improvement and measured delivery of agreed requirements (In Full On Time to A1 specification).

5 Establish a relationship that is based on a differentiated *do and improve* strategy against agreed baselines and other benchmarks, with a focus on total cost improvement, in addition to meeting or exceeding agreed product and service requirements In Full On Time to A1 specification.

6 Conduct regular business review meetings (for example, every six to 12 months) to review the relationship approach and performance, and identify opportunities for improvement.

7 Understand the degree of alignment between the relationship parties by sharing business goals, performance expectations, and relevant financial and non-financial information, as well as opportunities for improvement.

8 Engage high performing relationship managers and ensure they have clear roles and responsibilities as well as authority and empowerment levels for managing the relationship at a Major level. These people will include sales and project management staff, as well as asset management and procurement professionals who are respected for their competencies, knowledge and skills and have a proven commitment to quality and continuous improvement.

9 Consider a mix of fixed price, lump sum, fee for service, schedule of rates, unit pricing, time and materials, and incentive-based commercial arrangements linked to improvements against agreed baselines.

10 Implement a major account plan linked to KPIs and contractual or SLA obligations.

Type 7 Key relationships

Type 7 Key relationships, as pictured in figure 3.9, *do and add value* around agreed strategies. They are long-term, strategic relationships, exploiting synergies between customer(s) and supplier(s). All relevant information is shared to minimise areas of conflict and promote high levels of cooperation and innovation in order to add value and reduce total costs. In addition, these relationships aim to meet or exceed a complex set of agreed requirements IFOTA1 specification. Key relationships are outcomes focused, proactive, solutions selling, independent, preferred supplier based, win–win relationships that have multi-level interfaces.

Figure 3.9: Type 7 Key relationships

They are strategically important, complex and multi-dimensional in terms of products or integrated services, and can be linked into supporting, complex multi-level contract and contact management interfaces. Key relationships are focused more on genuine value for money performance criteria and value adding than on cost reductions. They are quality driven, continuously improving relationships, based on innovation. Both customer and supplier parties to the relationship provide expertise over and above the products and services given or received for which they are paid or pay. Organisations, departments, functions and teams use their strengths, skills and expertise to help the other party and to add value for the other customer(s) or supplier(s). This may involve:

- joint training, in areas such as problem solving, relationship management, safety and induction, and specific product and service training

- financial assistance and support, such as for accounts payable and receivable, and aggregate purchasing

- IT systems, in areas such as joint development, sharing and integration, and sharing business intelligence

- best practice, through the use of activities such as shared benchmarking information and work practices, and preventive, predictive or design-out maintenance techniques

- workplace reform and employee relations, in areas such as shared learnings and people exchange programs.

Key accounts are often referred to as enlightened self-interest relationships based on long-term win–win outcomes for all parties. An openly shared strategic key account plan is in place that drives relationship development and performance. This involves detailed sharing of business strategies and other relevant information, minimising areas of conflict and exploiting synergies between the relationship parties, and the use of compatible systems and processes. Key relationships may well be associated with multi-functional support groups or customer focus teams working within a complex, multi-level contact environment. However, management and responsibility for the relationship is still linked more to individuals and individual accountability than to teams and joint accountability, as would be the case in Partnering and Alliancing relationships.

Key relationships are the domain of outcomes-driven, professional and accountable key account managers, strategic supply and procurement managers, senior project and asset managers, business managers and mid to senior level relationship executives. Regular and formal business review and development meetings review progress against the documented key account strategy or plan, objectives and requirements, and to discuss future opportunities. While interdependence in the relationship is increasing, at its core the Key relationship remains an independent relationship.

Ten tips for implementing and sustaining a Type 7 Key relationship

Here are 10 ideas for engaging in a successful Type 7 Key relationship:

1 Agree on the value propositions that underpin the Key relationship, through identifying the cooperation benefits associated with a

preferred supplier engagement, exploiting synergies based on *do and add value* agreed strategies, solution selling, innovation and continuous improvement.

2   Exchange information early and often on strategy, vision and business objectives to ensure alignment of goals and opportunities for improvement. Identify and eliminate potential conflicts. For example, conduct a joint SWOT analysis to ensure there is a common understanding of the operating environment and opportunities for improvement.

3   Break down and analyse the relationship supply chain and internal value chain (primary and support activities). From this analysis agree on financial and non-financial baselines and identify value-adding opportunities, and people and change management opportunities. Focus on meeting or exceeding agreed product and service requirements.

4   Implement training and induction programs in the area of high performance relationship management and key account management to support the development of a Key relationship.

5   Develop, share and implement a Key relationship plan with agreed long-term objectives and milestones as the lead document in managing relationship development and performance. Support the plan by jointly agreeing on a relationship charter based on shared purpose or vision, relationship key objectives and guiding principles. This succinct one or two page document will be the basis of the moral agreement that underpins and directly supports any legally binding obligations or contractual requirements.

6   Conduct business review and development meetings every six to 12 months to review performance and identify opportunities for improvement.

7   Employ high performing, outcomes-driven Key relationship managers with clear, documented roles and responsibilities, and high levels of accountability and empowerment to lead and manage the relationship. Establish a lead team and support teams as required to steward the relationship approach and performance. The Key relationship manager is typically the team leader.

8   Consider the full range of commercial frameworks available, from fixed price and schedule of rates through to open book, shared risk–reward arrangements.

(continued)

Ten tips for implementing and sustaining a Type 7 Key relationship *(cont'd)*

9 Conduct facilitated workshops to support the innovation and improvement process, such as relationship review and improvement workshops.

10 Develop a highly adaptive, flexible, responsive, cooperative, customer-focused, performance-based culture and establish integrated systems and processes to manage a complex set of two-way requirements. Review internal relationships for willingness and capability to effectively support the Key relationship approach.

Partner segment relationships for relationship types 8, 9 and 10

The partner segment relationships offer greater scope for relationship development than the vendor (types 1 to 4) or supplier relationships (types 5 to 7). This does not in any way devalue the importance of vendor and supplier relationships, but rather suggests that there are other relationship choices that in the right circumstances have the potential to deliver greater value. Movement from vendor and supplier relationships to partner segment relationships means turning coordination and cooperation into collaboration; dependent and independent relationships into interdependent relationships; and an environment of competitive tension into one of creative tension.

Partner segment relationships refer to principle-centred, trust- and transparency-based Type 8 Partnering and Alliancing relationships; the breakthrough, bold and different Type 9 Pioneering relationships; and the extended supply chain Type 10 Community relationships that are linked by a common purpose and common goals for the common good.

These three relationship types are invested in next practices as well as current best practices. They live in a world of team-based creative tension versus competitive tension. Sharing everything—including ideas, risk, processes, people and benefits—within a highly flexible, empowering, innovative and collaborative business framework is the preferred partners' modus operandi. Relationship managers, sponsors and champions play a prominent leadership role in these relationships. They are often referred to as partnering or alliance managers, pioneers, change champions or community leaders.

Partnering and Alliancing, Pioneering and Community relationships share the same principles, concepts and many of the same practices, but their form, structure, level of complexity and ambiguity can differ significantly.

Type 8 Partnering and Alliancing relationships

Partnering and Alliancing relationships (see figure 3.10) are the collaborative development of successful, long-term, strategic relationships based on mutual trust, best practice, sustainable competitive advantage and benefits for all the partners. They create a separate and positive impact outside the partnership or alliance.

Figure 3.10: Type 8 Partnering and Alliancing relationships

Most innovation in the future will demand that historically adversarial relationships — (1) between many functions in the firm, (2) between labour and management, (3) between suppliers and the firm, (4) between the firm and its distributors/customers — be replaced by cooperative relations.

Establishing new relationships requires listening, creating a climate of respect and trust, and coming to understand the mutual benefits that will ensue if partnership relationships are firmly established.

Tom Peters[2]

Partnering and Alliancing relationships are a logical response to the globalisation of markets, increasingly intense competition, the need for faster innovation and the growing complexity of technology. These relationships are now seen as a legitimate fourth growth option for businesses alongside organic growth, acquisitions and mergers, and divestments. It makes good sense that connected people, departments, companies, and customers and suppliers, who don't have to compete with each other, should actually work with each other for some agreed common purpose. Past adversaries are becoming collaborative colleagues. That said, successful partnerships and alliances are often compared to successful fourth marriages: both are a triumph of hope over experience!

Type 8 Partnering and Alliancing relationships are based on trust and transparency around a shared vision, common goals and a joint strategy for mutual benefit. They are intensely collaborative, principle centred, interdependent, performance-driven relationships with integrated interfaces, work teams and processes. Partnering and Alliance relationships inhabit a world of seamless boundaries, frictionless commerce, shared risk–reward, performance-based remuneration and joint benchmarking. There is a shared governance of the relationship by joint leadership, and management and operational teams, who hold themselves mutually accountable for the wellbeing and success of the relationship. In these relationships, firms leverage core competencies for continuous and breakthrough improvement around a strategic scorecard of leading and lagging performance measures. Partnering and Alliancing relationships are not only strategic but are also seen as critical to the long-term wellbeing and success of the partners. They are often viewed as relationship role models and centres of excellence, and are used as vehicles for internal transformation. These relationships have a separate and positive impact outside the relationship itself and are used as points of reference and referral for new business and growth opportunities. All parties to the Partnering and Alliancing relationship have something fundamental to gain or lose from the success or failure of the relationship.

Type 8 Partnering and Alliancing relationships are, above all, about mutual trust. They are based on competence, character, interdependence, honesty and integrity in fair-minded and reasonable people working together in good faith as individuals and teams to achieve shared visions and common goals for mutual benefit. The win–lose options have been removed and the relationships are based on the management and effective allocation of risk rather than 100 per cent risk transfer. Risk is managed by those partners best able to do so, with shared accountability linked directly to a gainsharing and painsharing remuneration model. This is the transition stage between the old and the new worlds, moving from phase 1 to phase 2 on the development curve (see figure 1.12 on page 23) where paradigm shifts have changed from rough concepts to practical application.

Partnering and Alliancing relationships would normally involve one-on-one or simple cluster relationships more than the virtual or extended networks and supply chains that apply to Type 9 Pioneering and Type 10 Community relationships. This is the transition point from being a traditional customer and supplier to being a partner. The great leap forward has occurred. This is now a world of sharedness, not just sharing. There will be a formal relationship development process, a relationship charter, and a jointly developed and owned strategy or action plan in place. Cross-organisational leadership and operational and management teams, and not individuals, manage the relationship, with sponsorship, commitment and leadership provided by senior management and the executive teams. The application of the moral agreement has now taken prominence over, but not necessarily displaced, the traditional contractual agreement. In many cases they are one and the same.

The good news is that Partnering and Alliancing relationships are appropriate both internally and externally to the organisation. The collaborative shared vision and common goals approach is as applicable to traditional adversaries, such as management, employee, union relationships; operations and maintenance divisions; high impact project teams, as to critically and strategically important external customer, supplier and stakeholder relationships and major projects.

A true Partnering and Alliancing relationship, whether between internal departments or externally between customers and suppliers, positively influences and permeates the whole organisation, giving employees a greater sense of purpose, providing better than could be expected operational performance and return on investment, and acting as a benchmark and role model for customer satisfaction and competitive advantage generally.

Project Partnering and Alliancing

Project Partnering and Alliancing is based on the same principles and many of the same practices as Partnering and Alliancing relationships, but differs in scope and timeframe. Project-based alliances and partnerships are linked into a more narrow strategic, project scope and schedule, operational focus. As the name implies, the spotlight is on a project, such as the building of a hospital, road and rail infrastructure projects, defence projects for ships and aircraft, and large shutdowns or turnarounds on manufacturing, petrochemical, and oil and gas sites. Organisations often use Project Partnering and Alliancing, on either a single project or a series of projects or programs, as a stepping stone to strategic partnerships and alliances.

Project-based, as well as long-term strategic, Partnering and Alliancing can appear in all market segments in both the public and private sectors. These relationships include reciprocal trade, where the customer–supplier role is reversed; internal relationships between divisional, departmental, functional and operational units;

union–management relationships; relationships with competitors and co-suppliers; and extended relationships up and down the supply chain, involving clusters, networks and consortia groups.

The keys to success for good alliancing are a clear, simple, alliance agreement with good performance measures, strong positive leadership, a good (team-based) structure that integrates the partners, no duplication of activities, one system, sitting together, having a vision of a virtual company with common goals and high levels of trust. This has to happen, but it requires a leap of faith. For partnerships and alliances to be successful you have to love your partner's profit.

Alistair Tompkin, Former director of power generation, Hazelwood Power[3]

As with total quality, many organisations find the language, principles and concepts of Partnering and Alliancing compelling, but have great difficulty putting the ideas into effective practice. For most organisations these relationships require paradigm shifting. However, many so-called partnerships and alliances are nothing more than glorified, conventional, contractual relationships with a twist of cooperative rhetoric. Signing off on a partnering charter is next to useless if the executive teams do not actively support and commit to the principles, and the details are not delivered by an informed, competent, committed and empowered workforce.

Although commonsense in principle, the reality is that genuine Partnering and Alliancing is far from easy. Many organisations talk about implementing Partnering and Alliancing relationships but still tend to be Combative and Tribal, particularly in these times of cost reduction, restructuring, downsizing, outsourcing and divesting. This departmental command and control approach often restricts communication, destroys initiative and creativity, and leaves the organisation diminished in its capacity to relate effectively with the outside world. Interdependence, total trust and transparency, sharing risks and rewards around common goals and performance-based remuneration are just some of the mindset shifts that need to be practically embraced in Partnering and Alliancing relationships.

There are always simple examples in nature that can show us the way, if we are only willing to listen and understand. Sometimes referred to as V power, the Lessons from the Geese is nature's way of explaining high performance collaboration and how it works.

LESSONS FROM THE GEESE[4]

1   As each bird flaps its wings, it creates an 'uplift' for the bird following. By flying in a 'V' formation, the whole flock adds 71 per cent more flying range than if the bird flies alone.
    Lesson People who share a common direction and sense of community can get where they are going more quickly because they are travelling on the thrust of one another.

2   Whenever the goose falls out of formation, it suddenly feels the drag and resistance of trying to fly alone and quickly gets back into formation to take advantage of the 'lifting power' of the bird immediately in front.
    Lesson If we have as much sense as the goose, we will stay in formation with those who are headed in the way we wish to go.

3   When the head goose gets tired, it rotates back into the formation and another goose flies at the point position.
    Lesson It pays to take turns doing the hard tasks, and sharing leadership with people, as with geese — interdependent with each other.

4   The geese in formation honk from behind to encourage those up front to keep up their speed.
    Lesson We need to make sure our honking from behind is encouraging — not something less helpful.

5   When the goose gets sick or wounded or shot down, two geese drop out of formation and follow him down to help protect him. They stay with him until he is either able to fly or dies. Then they launch out on their own to join another formation or catch up with their own.
    Lesson If we have as much sense as the geese, we will stand together and help each other just like the geese.

Partnering and Alliancing relationships require a paradigm shift and not just a matter of doing the same things better. It is about a fundamental change in attitude, mindset, behaviour, practice and performance. In all our relationships, not just Partnering and Alliancing relationships, we would do well to act more like the geese.

Ten tips for implementing and sustaining a Type 8 Partnering and Alliancing relationship

Here are 10 ideas for engaging in a successful Type 8 Partnering and Alliancing relationship:

1 Confirm the relationship scope, value propositions and alignment with the broader corporate and organisational strategies. Align and seek compatibility in values. Conduct a joint SWOT analysis to build trust and ensure a common understanding of the operating environment and opportunities for improvement.

2 Agree and sign-off on the Partnering and Alliancing relationship charter, which includes statement of the shared vision, joint key objectives and guiding principles for the relationship. This document is the moral agreement that will underpin any legally binding contractual obligations. It is typically embedded in the Partnering and Alliancing agreement. But don't count on the contract: Partnering and Alliancing relationships will need to be adaptive and flexible.

3 Identify and develop internal and external relationship and change champions, innovators and other key influencers to challenge norms, prevailing paradigms and drive success. Develop relationship management and, specifically, Partnering and Alliancing as core competencies for the organisation. Include, for example the development of and succession planning for internal relationship facilitators, partnering facilitators and high performance relationship managers to ensure the relationships are enduring and successful beyond the life of key people.

4 Learn to love your partner's profit and your partner's success—continually strive for win–win outcomes.

5 Implement a performance-based improvement plan around joint ownership and governance, clear accountabilities, multi-level interfaces and networks, integrated structures and processes, and shared risk and reward.

6 Gain buy-in and effectively manage internal stakeholders, leaders, key influencers and other third parties to be advocates and supporters for the Partnering and Alliancing approach.

7   Measure your progress. Develop leading (means-based metrics) as well as lagging metrics (KPIs) to manage the relationship journey. Leading measures could include the number of new ideas, unnecessary disputes or escalation of issues and number of surprises (pleasant and unpleasant). Lagging measures involve profit, return on investment KPIs, and safety performance outcomes, such as the number of lost time injuries, environmental and social impact measures.

8   Know how the other partner operates, how they share information, make decisions and allocates resources. Understand each partners' organisational structure and internal relationships (quality and performance), day-to-day work practices, and their policies and procedures, culture and norms. This can be done through site visits, engagement and alignment workshops, and informal networks.

9   Reach agreement with all the partners that Partnering and Alliancing relationships are defined as principle-centred, performance-driven, collaborative relationships based on trust and transparency around common goals and shared risk for mutual benefit. They are led by fair-minded and reasonable people who act in good faith. Align culture, strategy, structure, process and people around these common goals.

10 Evaluate the willingness and capability of all parties to partner effectively, including your own organisation: know yourself as well as understanding others. Exploit strengths, leverage core competencies and respect differences. Learn from each other by listening, sharing, coaching, mentoring.

Type 9 Pioneering relationships

Type 9 Pioneering relationships (see figure 3.11, overleaf) capture those paradigm shifters and pioneers who dare to seek new relationship boundaries and break old rules. These are brave, bold and different relationships involving empowered, accountable leaders and teams consisting of people who are passionate, stubborn and often unreasonable in their expectations. Pioneering practitioners find new solutions to seemingly intractable and impossible problems. Breakthrough thinking coupled with intelligent risk taking is encouraged, delivering both continuous and breakthrough improvement. Pioneers typically work best within flat organisational structures and empowered modular teams and interfaces.

Some Pioneering relationships examples include transformational initiatives from the global to enterprise level, integrated project breakthrough teams, advanced multi-partner alliances, virtual enterprises, co-producer relationships, social enterprise initiatives, business and social entrepreneurship, competitor collaboration, and ground-breaking public–private sector relationships.

We typically associate breakthroughs and step changes with technology and sports. The internet and iPhone, new world records, faster times and greater heights have touched, fascinated and inspired us all. But pioneering breakthroughs and paradigm shifts also occur in personal and business relationships, and in the way we associate with each other. Combining the different generations (baby boomers, gen X, gen Y), the potential of the internet, the speed of technological development and the hardwired human desire to improve creates a ripe environment for Pioneering business relationships.

Figure 3.11: Type 9 Pioneering relationships

Pioneering is here to stay

The good thing about Pioneering is you don't have to keep up with the Joneses. Leapfrogging is not just allowed, it's preferred. Next practice is better than best practice. For example, many would argue Africa is at a distinct competitive advantage in communications because it has little investment in hardwired, copper-based telecommunications: in many cases these countries can leapfrog from nothing at all to mobile and wireless communication without being constrained by the legacy of old technology. The large-scale peer review of software development through collaborative open sourcing (such as Linux and Mozilla Firefox); the global collaboration required to build Wikipedia; the engaging communication applications and platforms that are Facebook, Google, YouTube and Twitter—all these online initiatives started as pioneering ideas, trials and development projects before they became communities and movements for social change.

Nine insights to better understand Pioneering relationships

- Pioneering relationships typically start small, often at a project level, grow and turn into role models.

- Pioneering relationships are often associated with eureka moments, those sudden, unexpected flashes of insight and discovery.

- Pioneering relationships have a mix of economic, social, cultural, environmental and personal impact. They can be life-changing experiences for the participants.

- The foundation of Pioneering relationships, how they get started and the solutions they implement are often counter-intuitive at the beginning—at first glance, they don't make sense.

- Pioneering relationships challenge the prevailing paradigms, the status quo, and therefore meet with resistance or mistrust: 'We don't do things that way around here'.

- Pioneering relationships involve some sort of breakthrough or fundamental change in attitude, behaviour, practice or technology.

- Pioneering is not without risk: there is no guarantee of success at a personal or business level.

- Pioneering and the associated breakthroughs and paradigm shifts are specific to time, place and circumstance. What was a breakthrough 20 years ago for one party can still have the same breakthrough effect today for another party. Alternatively, breakthroughs that occurred 10 years ago are common practice today.

- Pioneering relationships always involve pioneers, champions, inventors, provocateurs, business or social entrepreneurs of some sort who trigger, inspire and lead the change.

Such is the nature of the human spirit, that there will continue to be a place for Pioneering relationships. There will always be the need to update the Book of Firsts.

Ten tips for implementing and sustaining a Type 9 Pioneering relationship

Here are 10 ideas for engaging in a successful Type 9 Pioneering relationship:

1 Sign-off on a Pioneering relationship agreement that captures the spirit and intent of the relationship; it should be flexible, non-prescriptive, and responsive to creativity and innovation.

2 Develop a culture that supports calculated, educated and intelligent risk-taking with a 'who dares wins' mindset and attitude. Encourage people to think and act as entrepreneurs and empower them to solve problems and seize opportunities.

3 Create space and time for people to innovate and think outside the square in an atmosphere of creative tension that allows for respectful challenging of assumptions. Put your best people on innovation, and empower and challenge them to break through.

4 Set up a joint benchmarking forum with selected customers and suppliers (internal and external) to share information and develop best practices.

5 Develop a sound understanding of high performance relationship management principles and practices to support the Pioneering relationship with induction, training and personal development programs.

6 Ensure that the people who have to make the relationship work, at the operating levels, are informed, involved and committed to achieving the Pioneering relationship goals.

7 Establish a process of executive oversight, joint leadership and management teams and supporting integrated team structures with clear objectives and accountable KPIs, and clear roles and responsibilities to ensure effective stewardship. Ensure effective protection of intellectual property, patents or inventions.

8 Break down the Pioneering relationship strategy into manageable projects or business cases and assign owners or integrated project teams to manage them.

9 Develop a joint Pioneering relationship business plan around joint ownership and governance, clear accountabilities, KPIs, integrated processes, and shared risk and reward. Review the plan regularly.

10 Conduct facilitated workshops in support of the innovation and improvement process, such as relationship review and improvement workshops, process review and re-engineering workshops, technology review workshops, problem-solving workshops and requirements reviews.

The micro-credit story

Muhammad Yunus, founder of the Grameen Bank, was awarded the 2006 Nobel Peace Prize for his work in the area of micro-credit and micro-loans in Bangladesh. The journey towards improving the standard of living of the poor started with a loan of $27 of his own money to a group of Bangladeshi women. Extending very small loans to spur entrepreneurship to those in poverty is based fundamentally on trust and strong relationships. Yunus believed that, given the chance, poor people would repay the borrowed money and that micro-credit could be a viable business model. Extremely impoverished people, particularly women, have been able to start up self-employment projects as micro-entrepreneurs, generate an income and begin the process of building wealth and exiting poverty. Surprisingly, micro-credit was met with scepticism from the larger development organisations when first begun. At the end of 2009, the loan repayment rate, as measured by the ratio of the current performing to overdue loans, was greater than 97 per cent.[5] Micro-finance is recognised as having improved the lives of millions of people and has captured the attention of the traditional banking industry as a source of future growth. Ironically, one of the well-accepted solutions to eradicating world hunger is the education of women and the Grameen Bank, with the application of micro-credit, is directly helping and assisting this process.

The Bill and Melinda Gates Foundation story

The Bill Gates and Microsoft journey from a small suburban Californian garage to the world's largest software company is well documented as one of the great business success stories of the 20th century. Continuing the pioneering spirit, Bill Gates turned from being a successful business entrepreneur to being a successful social entrepreneur through the Bill and Melinda Gates Foundation. The foundation is the largest transparently operated, private foundation in the world: it had an endowment of $33 billion in December 2009 and donates more than $1.5 billion each year. Its primary aims are to improve healthcare and reduce extreme poverty around the world, and in America to expand educational opportunities and access to information technology. The manner in which the foundation applies business techniques to its operations, and its wealth and reach make it a pioneer and one of the leaders in philanthropic capitalism and global philanthropy.

Project Alcatraz

In Project Alcatraz,[6] a social entrepreneur champions social change through the application of business processes and entrepreneurial principles. Alberto Vollmer is heir to one of Venezuela's richest and oldest families. He is also a business leader and social entrepreneur who runs the family business, the Santa Teresa Rum Company, from a 200-year-old hacienda and sugar cane plantation in the state of Aragua. Aragua had one of the highest homicide rates in the country. Project Alcatraz commenced out of crisis in 2003, when local gang members mugged one of Vollmer's security guards and stole his gun. Vollmer needed to make a quick decision to avoid a potential escalation of violence with the gang. He caught up with the gang members and offered them two options: 'Either we go to the police and you go to jail or you work for us at the hacienda for three months.' To escape street life and the world of drugs, killing and violent crime, they chose the work option. Within two weeks the gang leader asked if he could bring more members along. In total, 22 gang members were involved in the first Project Alcatraz group.

Since then Vollmer and his company have taken 75 gang members, including members from the opposing gang, off the street and out of gang life, and some have gone on to start their own businesses. Vollmer has even enlisted a network of mothers to support the gang members and the project. The Santa Teresa Rum business is now profitable, safe and secure; many local gang members have been converted into productive members of society; and the evolving social entrepreneurship model has been recognised both nationally and internationally. In July 2009, the Alcatraz Project was awarded the best project for social inclusion at the Beyond Sport Summit in London.

The local community and the police were at first highly suspicious of this groundbreaking project and having such a high concentration of young criminals so close to town. The gang members themselves initially mistrusted Vollmer's motives. The results speak for themselves: over the last three years, serious crime in the local community has dropped by 76 per cent.

The project comprises three phases. First, three months are spent in the surrounding hills growing coffee. In phase two, which lasts for another three months, the gang members continue to work on the farm, getting counselling and playing rugby to learn teamwork and the benefit of shared goals, and working out any aggression. While there is a 15 per cent recidivism rate, for those who make it to phase three there is the prospect of gaining full-time employment at the Santa Teresa Rum Company. In many cases the lives of the gang members have been fundamentally changed for the better. They are taught the value of hard work, the value of money and how to be productive members of the local community.

This is a story of a pioneering project about change, principles and choices in turning a lose–lose point of crisis into a win–win lifetime opportunity.

Sydney Water operations and maintenance initiative story

For more than 15 years Sydney Water Corporation, Australia's largest water services supplier and statutory state-owned corporation, has been experimenting with different relationship approaches, ranging from traditional fixed price, lump sum, schedule of rates engagement to leading edge and innovative partnerships and alliances.[7]

(continued)

Sydney Water operations and maintenance initiative story *(cont'd)*

With total support and commitment from senior management, Sydney Water has embarked upon a groundbreaking change initiative to develop an internal partnering relationship between its operations and maintenance divisions, involving over 1500 people. This transformational initiative is having a positive impact well beyond these two divisions. As part of this initiative, by February 2010 Sydney Water had accredited and trained more than 20 employees as partnering facilitators to directly support and work with leaders and the operations and maintenance change process. Prior to this initiative commencing in late 2008, the partnering facilitator role would typically have been given to external consultants. The partnering initiative has been extended to engaging a wider group of leaders, and other internal and external stakeholders. This significant investment in time, people and money will have a positive long-term impact. The key business drivers are to improve internal working relationships, and to develop a smarter, more efficient and effective business, and a culture of continuous improvement around improved availability and reliability of Sydney Water assets.

ICI workplace reform story

In the mid 1980s and 1990s, ICI Australia was a pioneer in workplace reform. For many decades the petrochemical and industrial complex at Botany in Sydney was a magnet for major industrial disputation based on poor management and associated work practices, all linked to a debilitating overtime culture — overtime was seen as a fundamental entitlement and overtime rates were fiercely fought over and negotiated by union executives, employees and management. Safety (people), reliability (plant and equipment) and productivity (people and the business) were the victims in such a work place. This was a brutally adversarial and confrontational work environment that, in the worst case, had the potential to bring the business to its knees.

A small group of like-minded management and union shop stewards who were also ICI employees came together in the spirit of good faith and implicit trust to discover new paradigms and avert the impending crisis. They agreed on a set of guiding principles around a shared vision and common goals for a safer and more productive work environment.

As an employee of ICI Australia around this time and working as a manager in a variety of sales, market specialist and manufacturing roles, I was affected by two distinct paradigm shifts that came out of the changes that resulted. First was the move to treating and trusting employees and unions as strategic partners in the business and not mistrusting them as enemy combatants. This resulted in quite a remarkable change in attitude and behaviour. Second was the joint agreement struck between management, union and employees to move from an overtime environment to payment by annualised salary. The annualised salary was based just below what the employees were previously earning from a small base pay and large overtime component.

Surprisingly, this was not an easy sell. The benefits of annualised salary and workplace reform were not immediately obvious at all to many folk, from senior managers, to employees and the unions themselves. One senior management at head office said, 'Why should we pay these people their base plus overtime equivalent if we are expecting fewer hours to be worked?' What the manager and many others didn't appreciate was that over decades the employees, with management support, had built up a lifestyle and standard of living around overtime. Rather than rewarding improved reliability, availability and productivity, this overtime practice did the reverse. While the standard of living rose with more overtime, the workers' quality of life had deteriorated. Increased stress levels, poor work–life balance, unhappy families and increasing divorce rates were the result.

After the first weekend we went from overtime to annualised salary, we arrived to work on Monday morning and found that the plant uptime was virtually 100 per cent over the weekend and overtime worked was virtually zero. The win–win was seen by every company employee that day. Based largely on their own collective efforts and responsibilities, the unionised employees effectively had a choice as to whether they worked unpaid overtime or worked smarter in fewer hours for similar pay. In not working overtime, which in some cases meant going from 60–80 hours per week to less that 40 hours, but getting paid a similar amount through annualised salary, their quality of life improved significantly. Safety, reliability and productivity improved significantly. Management could not have been happier. A truly paradigm shifting win–win.

(continued)

ICI workplace reform story *(cont'd)*

In 1995, I was involved in the fifth generation enterprise bargaining agreement discussions at ICI, to produce a strategic partnership agreement. This was a life-changing experience for me, as I saw first hand the benefits that can be gained when people embrace positive change. Under the new agreement, workers within self-managed work teams were put in charge of day-to-day running of the petrochemical complex; union delegates became workplace facilitators; and site review teams comprising management and union representatives managed the results. Management then had more time to devote to larger strategic issues. The new enterprise agreement was hailed as breakthrough for cooperation in the workplace.

For many organisations today the Pioneering story is business as usual. For other organisations in many countries, this same story could happen today and still be a pioneering paradigm shift.

Type 10 Community relationships

Type 10 Community relationships (see figure 3.12) comprise individuals, groups, networks or organisations that work together for a common purpose, to achieve common goals for the common good of the community. Type 10 Community relationships include integrated supply chains and value chains, the extended enterprise, business cooperatives, social enterprises, open source communities, social media, community groups and organisations. They come in various shapes and sizes, ranging from supply chain communities in business to emergency response communities, indigenous enterprise communities and healthcare communities. Operating within the partner segment of the 0 to 10RM Matrix (figure 1.4 on page 10), Type 10 Community relationships exhibit many of the same principles and practices as Type 8 Partnering and Alliancing relationships and Type 9 Pioneering relationships. The difference being these principles and practices are extended beyond conventional organisational and social boundaries.

Type 10 Community relationships are legacy building and collaborative, delivering sustainable triple bottom line benefits (business, social and environmental). Communities are often multi-layered and complex. They can operate online and offline, in both the public and private sectors, and for profit or not-for-profit.

Figure 3.12: Type 10 Community relationships

Participants share a common sense of community and working democracy, sharing information, leveraging skills and competencies, pooling and sharing resources across the community. Community relationships are populated on the whole with selfless, inclusive, giving and caring people, who are open and outward looking, linked by an interconnected and shared destiny.

Internally, Community relationships present as informal employee networks or communities of practice involving connected and networked individuals, teams and groups, a style of organisation often far removed from the usual formal organisational chart of functions and departments.

As one of the world's leading automakers, Toyota's extended supply chain relationships with their suppliers are legendary and well documented, and these relationships have been a key element in their success for over three decades. The community spirit and approach is implicit and ingrained in Toyota's culture of continuous improvement and innovation, teaming and respect for people. Typical practices that extend out of this community mindset include early supplier involvement in design, joint innovation and development, integrated supplier managed inventory systems, Just In Time approach to deliveries, supplier co-location, building multiple tier relationships with suppliers, and building strong employee engagement, collaboration and loyalty.

Formed in 2001, New Zealand's Fonterra has 16 000 employees and is one of the world's largest and most successful dairy companies, responsible for more than one-third of the world's dairy trade. Fonterra exports 95 per cent of its product to customers and consumers in more than 140 countries.[8] Fonterra is also a business cooperative owned by 11 000 dairy farmers, who as shareholders, empower the Fonterra Board and executive team to run the company on their behalf. The strategy behind the cooperative business model is sustainable growth while building trusting customer partnerships on the lowest cost base through supply chain integration and innovation. In comprising over 95 per cent of New Zealand farmers in the manufacture and delivery of essential dairy products, there is a genuine community-based social, economic and environmental triple bottom line impact and focus.

Angel Flight Australia is a charity that provides free, non-emergency flights to people in both financial and medical difficulty. This is a classic win–win community effort around a common purpose of assisting patients and carers to travel to and from medical facilities anywhere in Australia. Angel Flight is often life-extending and life-saving for people living outside the major cities or in remote areas. Angel Flight Australia has all the qualities of a high performance Type 10 Community relationship. It is collaborative, win–win and legacy building, sustained by altruistic people who share a common purpose and common goals for the common good. It is a win in terms of getting effective healthcare to clients who are in poor health, financial difficulty and face daunting distances to travel. It is a win for the volunteer pilots, as they get to follow their passion for flying, while they are helping others who are less fortunate. It is also a win for healthcare professionals in optimising their time, efforts and value to the community.

Volunteer, non-profit communities and community organisations at the local and international level are also operating in Community relationships. Habitat for Humanity[9] is one such non-profit, volunteer labour and donor-based community organisation with global wealth and reach. It works to eliminate poverty housing and homelessness from the world by inviting volunteers from all backgrounds, races and religions to build houses together in partnership with families in need. Habitat for Humanity has built more than 400 000 houses around the world, providing more than 2 million people in 3000 communities with safe, decent, affordable shelter. Habitat houses are sold to partner families at no profit and financed with affordable loans. Independent, non-profit affiliates or community-level Habitat for Humanity offices act in partnership with and on behalf of Habitat for Humanity International.

Other examples of Type 10 Community relationships include the virtual communities we associate with the internet and online business environment. Open source IT communities, such as Linux, Apache HTTP and Mozilla Firefox are built on open and transparent source code sharing and free contributions from users and developers. This is a global, seamless network of like-minded people pooling and sharing knowledge for the benefit of their open source communities.

Since Wikipedia, the free content, community web-based, multilingual, encyclopedia started up in 2001, millions of collaborative volunteers, linked to a common purpose of free online content have created more than 13 million articles. The principles under which Wikipedia operates are closely aligned to the Type 10 Community relationship descriptors mentioned. Wikipedia's key operating principles include maintaining a neutral point of view; free online content that anyone can edit; Wikipedians respecting each other and acting in good faith; and in the pioneering spirit, there are no firm rules — community members are asked to be bold but not reckless.

The growing world of online social media communities, through vehicles such as blogs, wikis, Twitter, YouTube and Facebook, is greatly impacting traditional offline business models (for example, print media) as well as online business communication and collaboration.

The business world and society at large are just starting to understand the power of community and are beginning to exploit the benefits associated with networked collaboration at a local and global level. This is particularly so in a world of such rapid and dramatic, online and offline, political, economic and social change. Type 10 Community relationships, although the last to be discussed on the 0 to 10RM spectrum, probably harness the greatest potential for individuals and organisations to make a difference. The future is a very exciting place indeed.

Ten tips for implementing and sustaining a Type 10 Community relationship

Here are 10 ideas for engaging in a successful Type 10 Community relationship:

1 Agree on the value propositions, goals and deliverables that underpin the Community relationship. These are typically based on collaboration benefits, triple bottom line benefits (business, social and environmental) and leveraging of core competencies across the community and associated stakeholders for mutual benefit.

(continued)

Ten tips for implementing and sustaining a Type 10 Community relationship *(cont'd)*

2   Establish long-term relationships that are inherently collaborative and legacy building, with high expectations.

3   Develop strong, formal relationships with key leaders and senior management sponsors. Identify and empower leaders and key influencers who are altruistic, open and outward looking, and linked by the interconnected and shared destiny that is the fabric of the community. Have in place a strong governance structure; clear, focused goals, responsibilities and accountabilities; and open, honest, seamless communication.

4   Establish agreements and commercial frameworks that are flexible, and adaptive around managing the complexity of community interests. The agreement will capture the spirit and intent of the Community relationship in addition to any contractual or legally binding obligations.

5   Establish a relationship charter detailing the shared vision, joint key objectives, rules of engagement and guiding principles for the relationship. Typically a one or two page document, it will provide clarity, purpose and direction for all Community participants.

6   Have in place a multi-level, integrated KPI scorecard to manage performance against community expectations and agreed objectives.

7   Establish open and democratic structures, with clear leadership roles, responsibilities and collective accountabilities to manage the ever-changing, complex multi-level interfaces that exist at an individual, team, work group and organisational level.

8   Community leaders will have to walk a fine line around the need to be authoritative in setting clear standards and direction; democratic in seeking viewpoints and in their decision making; and affiliative in showing empathy and support to community members and others.

9   Identify and develop internal innovators, champions and key influencers to support leaders, effectively engage the broader community, and directly assist in the delivery of community goals and objectives. Ensure all people at the operating levels who have to make the relationship work are informed, involved and committed to the community approach and the associated common goals.

10 Make time available for effective community participation. For example conduct site visits across the community to meet people, understand the culture or members of the community, share information and best practice, cross-fertilise ideas, resolve issues, build trust and develop friendships. Consider people exchange programs and secondments to increase skills and learning.

Summary

- The horizontal axis of the 0 to 10RM Matrix comprises the three relationship segments and 11 relationship types.

- The three segments are:
  - *the vendor segment* (Type 1 Combative, Type 2 Tribal, Type 3 Trading, Type 4 Transactional relationships), which has a *buy and sell*, coordination focus
  - *the supplier segment* (Type 5 Basic, Type 6 Major, Type 7 Key relationships) which has a supply and procurement, cooperation focus
  - *the partner segment* (Type 8 Partnering and Alliancing, Type 9 Pioneering, Type 10 Community relationships), which has a shared risk and benefit, collaboration focus.

- The 11 relationship types can be summarised as follows:
  - *Type Zero relationships* — non-existent, indirect or aspirational relationships that sit outside of the vendor, supplier and partner segments
  - *Type 1 Combative relationships* — adversarial, aggressive and confrontational
  - *Type 2 Tribal relationships* — those that defend and protect information, position and power base
  - *Type 3 Trading relationships* — opportunistic, deal-based relationships that have limited loyalty and differentiation
  - *Type 4 Transactional relationships* — arms length, impersonal, technology driven
  - *Type 5 Basic relationships* — *do and charge* against agreed scope
  - *Type 6 Major relationships* — *do and improve* against agreed baselines
  - *Type 7 Key relationships* — *do and add value* against agreed strategies
  - *Type 8 Partnering and Alliancing relationships* — based on trust and transparency around common goals for mutual benefit

- *Type 9 Pioneering relationships* — brave, bold and different, and paradigm shifting
- *Type 10 Community relationships* — essentially altruistic relationships that have a common purpose, around common goals for the common good
- All relationship types are legitimate. The strategic and commercial value underpinning the relationship and the willingness and capability of the parties in the relationship will determine which relationship type is appropriate to the circumstances.

CHAPTER 4
KEY COMPONENTS

The devil hides in the details—prowling for gaps, inconsistencies, loopholes, misinformation and mistrust, all to make mischief with. We need to meet the devil head on and get the components right, develop trust and build a sustainable platform for delivering high performance. In managing the details, high performance relationships need to go beyond the simple sales pitch, the time-tested ritual of offer and acceptance, and often drawn-out, detailed negotiations. To survive, endure and grow beyond the initial product and service offering and to avoid over-promising and under-delivering we need to implement and manage details or components. Even for traditional Type 3 Trading, Type 4 Transactional and Type 5 Basic buy and sell relationships, opening, probing, supporting and closing around immediate customer needs will not be good enough to sustain high performance. Understanding requirements (current and future), matching features to benefits, and handling indifference, scepticism and misunderstanding are all important professional selling and procurement skills. However, additional competencies must be employed to sustain high performance and to ensure that the relationship thrives beyond the life of key people.

Typically, the most creative and collaborative time in the relationship development journey or life cycle is at the go to market bid development and delivery stage, or at a point of crisis where there is little to lose and everything to gain. The participants in the process are empowered with seemingly unfettered freedom to be innovative and creative. How does one create an environment where that creativity and energy is ongoing?

The key components of 0 to 10RM lie at the heart of identifying and implementing the details required to make the relationship work. Key components are customer centric, as implied by Principle 3, Customers are the reason suppliers exist. The key components theme is shown diagrammatically in figure 4.1 (overleaf).

Figure 4.1: key components of 0 to 10RM

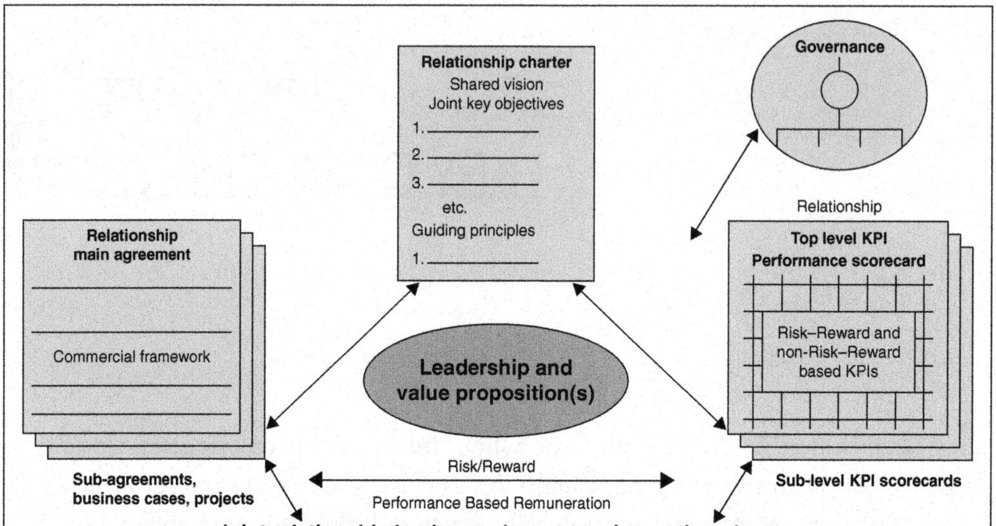

Joint relationship business plan, strategies, action plans

Key objectives	Initiatives	Who	By when
1.	1.1 1.2		
2.	2.1 2.2		
	etc.		

Note: KPI = Key performance indicator

The 0 to 10RM key components provide an insight into the detail that underpins high performance relationships wherever they may lie on the 0 to 10RM Matrix. They provide the link between the strategy of the organisation and the delivery of results. In summary, the key components are:

- *Leadership*—leaders are required to establish and maintain a relationship; be willing and capable; demonstrate courage, intuition and a long-term perspective to envision compelling alternatives to the current state; lead the process of change; engage, enable and empower people to effectively participate in the relationship management journey. Leadership is not only about executive commitment, but also leadership and ownership at all levels.
- *Value propositions*—to outline benefits and opportunities beyond a cheap price or low cost.
- *Legal contracts and agreements*—to document the relationship, and its scope, requirements and legally binding terms and conditions.

- *The relationship charter* — to document the vision or purpose, key objectives and guiding principles and behaviours of the relationship; it acts as the basis of the moral agreement for the relationship.

- *Relationship key performance indicator (KPI) and performance scorecards* — to form the basis for aligning metrics, performance measurement, risk management, relationship management, behaviours, continuous improvement and remuneration.

- *Risk–reward performance-based remuneration models* — to build trust and transparency and allow both risk and reward to be shared and managed.

- *Relationship plans* — to provide the roadmap for relationship improvement.

- *Governance* — to define the nature of stewardship of the relationship and its supporting structures, roles and responsibilities.

Each of these components has a different meaning and application depending on the relationship type or approach in place. Table 4.1 (overleaf) provides a summary table of descriptors associated with each component across each of the 10 relationship types, Type 1 Combative to Type 10 Community.

It is not the intention in this book to look at each component in detail across every relationship type. Each of the components is a special interest area in itself, around which entire books and training programs have been built. Instead, this chapter will give an overview of the key components to be considered when engaging in high performance relationship management.

Leadership and value propositions

For all the leaders out there, here is a key question: Why would anyone want to be led by you?

Rightly, leadership and value propositions are at the centre of the key components theme in 0 to 10RM. While the priority and importance of each component in a relationship will depend on circumstance, leadership and value propositions are, at the very least, first among equals. Leadership here has a broad interpretation and includes senior level and executive leaders with a vision, a purpose and a plan; mid level leaders who manage the top-down intent and the bottom-up engagement; and shop floor leaders who are empowered and committed to drive the delivery of the plan. All of them taking ownership around their clear roles and responsibilities.

Table 4.1: key components of 0 to 10RM

	Type 1 Combative	Type 2 Tribal	Type 3 Trading	Type 4 Transactional	Type 5 Basic	Type 6 Major	Type 7 Key	Type 8 Partnering and Alliancing	Type 9 Pioneering	Type 10 Community
Leadership	Directive Command and control Untrusting Arrogant Adversarial Coercive	Defensive Protective Authoritative Affiliative within the tribe (brother's keeper) Self-interest focus	Opportunistic Work hard and play hard negotiators Autocratic Pacesetting deal makers and traders	Authoritative Task driven Impersonal Faceless Arms length Service oriented	Task focus IFOTA1 focus Authoritative around standards Pacesetting around delivery	Results driven Outputs focus Authoritative Quality, service and continuous improvement based values	Consultative Democratic Authoritative Solution driven Team players	Collaborative Performance driven Principled, Authoritative Coach, democratic Affiliative	Passionate, stubborn unreasonable in their demands Pacesetters Paradigm pioneers, trail blazers	Authoritative in setting standards democratic in decision making Affiliative re empathy and support Selfless and caring
Value propositions	Win–lose outcome Financial focus Tough and aggressive Brand image Dollar and market share More dominant competitive position	Protect and defend information and knowledge, dollar and market share Intellectual property position and power base	Low or least cost Cheaper prices Undifferentiated Commodity focus Deal based Low switching costs Competitive margins	Faster technology Systems uptime Richness and reach Improved speed and efficiency Lower transaction costs	Ease of access Simple *do and charge* Greater convenience IFOTA1+ performance at competitive cost Repeat business and less tendering	*Do and improve* against agreed baselines Outsource non-core activities IFOTA1 focus on reducing cost base	*Do and add value* around agreed strategies Exploit synergies Innovation opportunities Cooperation benefits	Interdependence Win–win Collaboration benefits - value add, total cost - people, technology - brand/reputation - integration benefits	Stretch and breakthrough innovation Accelerated improvement Paradigm shift(s) Calculated risk taking	Triple bottom line based (social, economic and environmental) and legacy benefits Common purpose for the common good
Contracts agreements internal SLAs	Legalistic Tightly managed One sided Risk transferred Hard nosed dollar Punitive	Defensive and protective Ownership focus Control focus Restricted information sharing and communications	Deal based terms and conditions Cost and schedule focus Undifferentiated Standard quality specifications	Systems driven and automated Pay and go focus Standard terms and conditions rule Little or no negotiation involved	*Do and charge* on agreed scope Cost plus FFS Prescriptive, inputs IFOTA1 focus, terms and conditions	Outsourced scope linked to KPI measures and contractual performance obligations and terms and conditions	Complex multi-level contract and contact management interfaces Outcomes focus Complex requirements	Performance based Principle centred Trust focus, flexible Joint strategy Common goals Sharing risk and reward Mutual benefit	Flexible and adaptive to creativity/innovation Leveraging core competencies Intellectual property a key element	Flexible and adaptive around managing complexity of community interests Open, transparent documentation

Relationship charters	Not appropriate Not encouraged Too open and transparent for Combative relationships	Used to reinforce Tribal or group loyalty Mainly internal application WIIFM based	More company slogans or statements of customer service than charters Buy and sell focus	More generic, broad market statements of values, and service level commitments	Often implied or inferred or assumed by return or repeat business, i.e. integrity based, typically not documented, 'do what you say'	Charter is results driven customer-centric with an improvement focus on cost, quality, schedule, service levels	A moral agreement to support legal contract Strategic in intent Not legally binding A performance management tool	A strategic engagement document linking purpose, performance, measurement and attitudes	Frame the vision/ purpose Simplify complexity Provide clarity, direction and rules of engagement	Public documents for communications, engagement, public debate and review A moral agreement with broad community support
KPI scorecard performance measurement	Win/lose KPIs Short-term profit and financial focus Tough and aggressive One sided	Defensive KPIs for protecting position and self-interest Used for leverage, asserting blame or finger pointing	Simple cost, schedule and quality KPIs	Simple service level KPIs based on what, who, when Trackable, traceable Efficiency, speed, systems focus	IFOTA1 input focus. KPIs based on quality, cost, schedule, safety Formal and informal reviews of product and service delivery	Output KPI focus on short to medium term cost, quality, schedule, service levels and performance trends opposite baselines	KPI focus on medium to long term, strategic financial and non financial outcomes, innovation and process improvement	Strategic KPI scorecard approach including Legacy focus Direct link to charter and risk-reward sharing	A few simple KPIs frame the outcomes Stretch, breakthrough and legacy focus Linked primarily to improvement and innovation	Financial and non-financial KPIs representing community and legacy outcomes rather than vested interest outcomes
Risk–reward Performance-based remuneration	Risk transferred fixed price, rates fees, fees focus Non-transparent Short term profit and financial focus	Not transparent, risk averse or risk linked to fixed price, fees, rates schedule	Risk and reward separated and not shared More downside penalties than upside rewards	By exception only Pay and use focus Standard terms and conditions cover risk Fixed price or fee includes reward or margin	Typically done via return or repeat business based on performance and competitive, profitable costs and fees	Linked to KPI performance against baselines or benchmarks Focus on direct costs and total cost of ownership	Direct risk and reward managed and measured at tactical or operational and strategic levels Innovation focus	Open, transparent shared risk and reward based on overperformance or underperformance against agreed KPIs Operational, strategic and legacy based	Risk-reward inherent in the breakthroughs required Pioneers often bet their future around unclear risk profiles	Open, transparent shared risk and reward Individual responsibility Collective accountability

(continued)

Table 4.1 *(cont'd)*: key components of 0 to 10RM

	Type 1 Combative	Type 2 Tribal	Type 3 Trading	Type 4 Transactional	Type 5 Basic	Type 6 Major	Type 7 Key	Type 8 Partnering and Alliancing	Type 9 Pioneering	Type 10 Community
Relationship plans	Separate strategies/ action plans. Tightly managed and controlled by each party	Separate strategies or action plans. Demarcations. Secretive and controlled	Simple action plans based on what, who, when. Trackable, traceable, deal based. Efficiency and effectiveness focus	Systems, process, technology driven and automated. Depersonalised	Basic relationship management plans, focus on simple what, who and when actions or plans	Major account plans linked to KPI measures and contractual obligations and supply chain analysis	Strategic Key relationship plan leads relationship development and performance	Joint relationship business plan. Flexible and adaptive. Performance and improvement based. Integrated processes	Joint business/ action plans built on the sense of urgency and breakthroughs required	Community based business plan. Triple bottom line and legacy focus. Common purpose. Common goals
Governance	Command and control. Hierarchical. Hostile interfaces and abrasive rub points	Territorial interfaces. Informal contacts and networks. Backroom deals. Not open	Simple or single point of contact interfaces. Face to face or electronic	Electronic or single point of contact. Arms length. Impersonal	Simple, single or limited points of contact. Work to rule. Review is face to face or electronic	Regular formal performance reviews at mid-senior levels. Medium-high level contract management interface or focus	Work/project teams directed by key relationship managers. Stewardship from senior managers. Regular BRADs	Joint ownership, accountability, one-team approach. Joint leadership and management teams	Pioneers, paradigm shifters tend to be dominant players within small, tight governance team based structure(s)	Extended supply chain and community governance. Open source type governance structures

Notes: SLAs means service level agreements; BRADs = business review and development meetings; FFS = fee for service; IFOTA1 = In Full On Time to A1 specification; WIIFM = what's in it for me

There is a difference between leadership and management. Leadership is of the spirit, compounded of personality, vision and training. Its practice is an art. Management is a science and of the mind. Managers are necessary, leaders are indispensible.

Admiral J Moorer, US Navy[1]

The role and importance of leaders cannot be overstated. You can't start your journey unless you know where you are going. This is leadership from the bottom up as well as top down in the organisation, and it involves both strategy and tactics. However, as important as the groundswell of support is, without executive commitment and stewardship, the relationship improvement initiative, with rare exception, is doomed to fail. Ground floor and mid level relationship change champions can run a stealth, under the radar, campaign for only so long. This proceed-until-apprehended approach requires, at the very least, senior management endorsement and support. Remember, employee self-sacrifice is not a sustainable resource!

Leaders, by their action or inaction, can make or break relationships, which has a corresponding impact on people. Leaders set the vision and broad strategies for the organisation, in line with personal and company values, the external environment, and the organisation's internal support structure, capabilities and potential. Through their support, commitment, active listening and participation, leadership styles, vision and inspiration, leaders create an environment that makes the relationship happen. While there are many roles for managers and leaders to play, I see the following roles as key to successful high performance relationship management and in particular, Type 8 Partnering and Alliancing, Type 9 Pioneering and Type 10 Community relationships. Good leaders need to:

- Set and communicate an achievable vision, a broad strategy and a set of principles and standards that inspire everyone in the organisation, from the CEO to people on the shop floor. Effective leaders have a genuine long-term focus.

- Lead and coach others in creating a learning environment based on continuously improving skills and competencies, on the delivery of superior performance, and on trust — an environment in which individuals and teams take ownership and are encouraged to perform, and are empowered and committed to their full potential.

- Suitably reward and recognise high performance and outstanding achievement in skills and competency development, innovation and leadership, whether individual or team based. This is not about a twice-monthly salary slip or annual bonus, but rather about creative and flexible remuneration that generates loyalty to the organisation, personal satisfaction and work–life balance for employees and their families.

- Create an environment where people enjoy coming to work — a place where they are challenged by the expectations and the opportunities; where initiative is encouraged, and honourable, courageous and intelligent failure is rewarded; and where the barriers and obstacles to effective communication and performance are removed.

- Support and actively participate in the relationship management and improvement process.

Leadership styles vary and will differ depending on the relationship approach being taken for both the current state and desired future state. Daniel Goleman's six leadership styles[2] and his work on emotional intelligence have a direct relevance for 0 to 10RM. Coercive, authoritative, affiliative, democratic, pacesetting and coaching leadership styles are simple to understand and align well with the 0 to 10RM relationship types. Table 4.1 (on pages 120 to 122) gives a guide to which styles are relevant in each relationship type, from Combative to Community.

Just as there are different relationship approaches, no single leadership style should be relied upon exclusively. Just as we choose the right golf club depending on the lie of the ball, the distance from the hole and the prevailing conditions, so we will choose the most appropriate leadership style based on the current and future relationship state, and the prevailing conditions at a strategic and tactical level.

High performance relationship management needs to be underpinned by a strategy and a set of value propositions that support the nature of the relationship type and performance level engagement. A set of business drivers is required for the kind of relationship you want to sustain.

Value propositions are the underpinning benefits and opportunities that go beyond a low cost and a cheap price. They are the reason the relationship exists in the first place. This statement links directly to Principle 3, Customers are the reason suppliers exist. Some examples of value propositions as they apply to high performance relationships, especially Type 6 Major, Type 7 Key relationships, Type 8 Partnering and Alliancing, include:

- easier and faster access and transition to new technologies

- the ability to leverage off or have access to global reach, capabilities, brand, market knowledge and networks

- removal of complexity and reduction in total operating costs

- improved speed to market for new products and services

- increased revenues, additional profits, increased margin through risk–reward linked, performance-based remuneration

- greater integration, simplification and efficiency of operations

- improved brand and reputation

- significant improvement in risk management or significantly reduced or improved risk profile(s).

The nature of value propositions will vary from left to right in going from Type 1 Combative to Type 10 Community relationships, as shown in table 4.2 (overleaf). Value propositions will be representative of the strategic value and commercial value of the products and services associated with the relationship.

Table 4.2 (overleaf) summarises the high level generic value propositions and leadership styles associated with each relationship type. They can be adapted, contextualised and modified to suit the circumstances within the operating and work environment.

Contracts and agreements

Different relationship types attract different styles, content, terms and conditions in contracts and agreements. Different negotiation styles and techniques will also come into play depending on the relationship approach and performance level. Typically, external contracts and agreements are legally binding documents that customers and suppliers, clients and service providers enter into to enable the scope, product and service delivery of a job to be completed IFOTA1 specification. On one side, delivering value for money for the customer, and on the other side gaining a fair and reasonable return on investment (ROI) for the supplier. Internally, between business units, functions, departments and work teams, contracts and agreements often present themselves as service level agreements (SLAs) or similar. Again, the SLAs have a similar style, structure and content to external agreements and contracts depending on the nature of the relationship in place. In simple terms, and irrespective of internal or external application, contracts and agreements typically cover work scope (product or service), requirements (for example, quality, service, support, risk management), price, and terms and conditions.

Table 4.2: summary of associated value propositions and leadership styles and characteristics across the 10 relationship types

Relationship type	Typical value propositions underpinning relationship types	Leadership styles and characteristics
1 Combative	Win–lose outcome; better financial focus and outcomes; tough and aggressive presence; enhanced brand image; financial and market share improvement; more dominant competitive position	Directive; command and control; untrusting; arrogant; adversarial; coercive
2 Tribal	Protect and defend information and knowledge, financial and market share; improved financial and market share; more secure intellectual property position and power base	Defensive; protective; authoritative; affiliative within the tribe (my brother's keeper); self-interest focus
3 Trading	Low or least cost strategy; cheaper prices; undifferentiated products and services; commodity focus; deal-based benefits; low switching costs; competitive margins	Opportunist; work hard and play hard negotiators; autocratic; pacesetting dealmakers and traders
4 Transactional	Faster technology; systems uptime improvement; greater richness and reach for products and services; improved speed and efficiency; lower transaction costs	Authoritative; task driven; impersonal; faceless; arms length; service oriented
5 Basic	Ease of access; simple *do and charge* relationship; greater convenience and efficiency; IFOTA1 focus plus performance at competitive cost; repeat business and less tendering, lower costs	Task focus; IFOTA1 focus; authoritative around standards; pacesetting around delivery
6 Major	*Do and improve* against agreed baselines; outsource non-core activities; IFOTA1 focus on reducing cost base	Results driven; outputs focus; authoritative; values based on quality, service and continuous improvement
7 Key	*Do and add* value around agreed strategies; exploit synergies; innovation opportunities; cooperation benefits	Consultative; democratic; authoritative; solutions driven; team player
8 Partnering and Alliancing	Interdependence; win–win, mutual benefits; collaboration benefits for value adding, cost improvement, people, technology, brand and reputation, and integration benefits	Collaborative; performance driven; principled; authoritative; coach; democratic; affiliative
9 Pioneering	Stretch and breakthrough innovation; accelerated improvement; paradigm shift benefits; calculated or educated risk-taking benefits	Passionate, stubborn, unreasonable in their demands; pacesetters; paradigm pioneers, trail blazers
10 Community	Based on triple bottom line (social, economic and environmental) and legacy benefits; common purpose for the common good	Authoritative in setting standards, democratic in decision making; affiliative regarding empathy and support; altruistic—selfless and caring

Table 4.3 reviews the style and characteristics of contracts and agreements for each of the 10 relationship types.

Table 4.3: style and characteristics of contracts and agreements across the 10 relationship types

Relationship type	Style and characteristics
1 Combative	Legalistic, tightly managed, one-sided, risk transferred, hard nosed, hard dollar; punitive downside for poor performance, little upside for better than target performance
2 Tribal	Defensive and protective, ownership focus, control focus; restricted information sharing and communications
3 Trading	Deal-based terms and conditions, cost and schedule focus, undifferentiated or standardised quality specifications
4 Transactional	Systems driven and automated, *pay and go* focus; standard terms and conditions rule; little or no negotiation involved
5 Basic	*Do and charge* on agreed scope, cost plus margin, fee-for-service based, prescriptive, input driven, IFOTA1 focused, company standard, terms and conditions
6 Major	Outsourced scope linked to key performance indicator (KPI) measures, baseline improvement, contractual performance obligations and associated terms and conditions
7 Key	Complex multi-level contract and contact management interfaces, outcomes focus against a complex set of performance requirements
8 Partnering and Alliancing	Performance-based, principle-centred, trust focused, flexible, joint strategy, common goals, sharing risk and reward, mutual benefit based plain English–style documents
9 Pioneering	Flexible and adaptive to creativity and innovation; leveraging core competencies; ownership and use of intellectual property form a key element
10 Community	Flexible and adaptive around managing complexity of community interests; open, transparent documentation

The simplest way to appreciate the differences between the style and content of contracts and agreements associated with each of the 10 relationship types is to start with almost any organisation's standard terms and conditions. These are typically Type 4 Transactional or Type 5 Basic in style, layout, intent and content. Download any software off the internet or buy products or services online, and before you pay and receive your purchase you will almost certainly be asked to confirm, or not, your agreement to a standard, non-negotiable set of commercial terms and conditions. Purchase everyday, general goods and services offline, and before payment is made, there will be an offer or acceptance of standard terms and conditions or they will be explicitly stated on the back of invoice statements.

Standard terms and conditions are typically made up of what is called boiler plate clauses and, depending on the market segment, these standard terms and conditions will include clauses covering such things as provision of services, payment fees and terms, intellectual property rights and responsibilities, insurances, warranties, indemnities, confidentiality, termination, dispute resolution, assignment and notices. The list is almost limitless, depending on the nature and scope of the products bought or services being engaged.

For the other relationships in the vendor segment, standard terms and conditions become more deal-based for Type 3 Trading relationships; more defensive and protective, ownership and control focused in the case of Type 2 Tribal relationships; and more legalistic, hard-nosed and punitive for Type 1 Combative relationships. Combative, Tribal and Trading contracts will be very different in the way they are negotiated, as well as in their final style and content, from those of Type 8 Partnering and Alliancing agreements.

Alternatively, in going further to the right on the 0 to 10RM scale, from Type 4 Transactional and Type 5 Basic relationships, contracts and agreements focus on more open, results-driven outcomes and improvement, offering greater flexibility, shared information, risk and reward, cooperation and collaboration. Features that distinguish Type 8 Partnering and Alliancing agreements from more traditional commercial contracts include the following:

- Partnering and Alliancing agreements are plain-English style, positive documents that state upfront what the value propositions and principles underpinning the relationship, and therefore the contract, are.

- The fair-dealing spirit, principles and intent of Partnering and Alliancing, and the vision and objectives for the relationship, are documented upfront.

- Consistent with a no-blame philosophy and the principle of interdependence in a Partnering and Alliancing relationship, a 'no dispute' clause is sometimes incorporated into the agreement. This involves waiving rights to litigation between the partners except in the case of wilful default. No liquidated damages of any sort will be paid. True interdependence is inbuilt, removing the win–lose option from the relationship.

- The executive steering board, which can have various titles (for example, Partnering and Alliancing Leadership Team, Project Alliance Board), is the ultimate review and management body for the relationship.

- The roles and responsibilities of the Partnering and Alliancing leadership and management teams are clearly defined.

- A team approach and shared financial motivation to achieve common goals are clearly documented.

- Performance measurement processes are clearly defined — for example, in the form of KPI balanced scorecards or performance scorecards.

- Appropriate risk allocation and risk-sharing mechanisms are documented and linked to gainsharing–painsharing, and risk–reward performance based remuneration models or mechanisms.

- A fast and effective issue resolution process is in place to ensure a win–win approach and outcomes in the event of conflict arising.

- Partnering and Alliancing agreements include a process and incentives for identifying and achieving breakthrough and continuous improvement or exceptional performance.

As indicated in the key components image in figure 4.1 on page 118, for large and complex relationships there may also be master agreements capturing the high level strategic intent and scope, and general terms and conditions, with separate legally binding agreements for individual business cases or projects. These individual project or business case agreements would then survive the master agreement in the event of termination or disengagement.

Appendix A provides an example of a master negotiation terms sheet, which details negotiation styles, and terms and conditions. It is typically associated with the development and negotiation of Type 7 Key, and Type 8 Partnering and Alliancing agreements. This terms sheet is an example of the detail to be discussed for inclusion in agreements and contracts. For further reading on Partnering and Alliancing agreements, refer to the *Strategic Partnering Handbook*.[3]

Relationship charters

The relationship charter is a moral agreement — it is not meant to be legally binding. It states the shared vision or purpose for the relationship, the key objectives to be achieved and the guiding principles or behaviours to be engaged. Relationship charters can appear across all relationship types, but are more typically associated with Type 6 Major to Type 10 Community relationships. They are particularly relevant to Type 7 Key, Type 8 Partnering and Alliancing, and Type 9 Pioneering relationships.

The relationship charter is the documented handshake underpinning the trust and intent between the relationship parties. In the right hands for the right reasons, it is a living and powerful document.

Relationship charter template

Company A logo   Company B logo

Shared vision or purpose

The shared vision or purpose statement articulates an understandable, believable, achievable, rich picture of the future that embodies values and aspirations and has wholeness and integrity. An example would be: Role model partners in delivering world-class results, products and services.

Critical success factors or key objectives

You may list six to 10 critical success factors or key objectives. These articulate a response to the question: What is the relationship expecting to achieve? Some examples may include the following:

1 Safety
- Zero harm to people, plant and environment

2 Requirements
- Meet or exceed agreed customer requirements and expectations
- Achieve specific project requirements on quality, cost and schedule

3 Sustainability and environment
- Meet or exceed stakeholder social and environmental expectations

4 Financial success
- Achieve an outstanding commercial outcome for all the partners
- Demonstrate that superior value for money is being achieved

5 Innovation and continuous improvement
- Deliver breakthrough and continuous improvement through innovation and joint benchmarking
- Eliminate duplication and waste

6 People and teams
- Empower teams and individuals to achieve their full potential

7 Relationship and communication
- Build strong, positive working relationships based on trust, respect and open communication
- Promote the relationship to achieve broader business outcomes

The key objectives are SMART (specific, measurable, achievable, relevant and trackable or timely).

Guiding Principles

These are fundamental or self-evident truths, mindsets, behaviours, beliefs, values and attitudes applicable to all relationship participants.

Typically eight to 12 principles are included. Some examples are:

- We value safety above all else.

- We act in a way that is best for the relationship.

- We act with integrity—do what you say.

- We engage straight talk—talk before you write.

- We practice proactive problem solving and joint resolution of issues.

- We value a no surprises approach to all aspects of the relationship.

- We value open, honest, timely, accurate communication and information sharing.

- We act with trust, probity and professionalism.

- We will be fair and reasonable and act in good faith.

- We reward and recognise people and their achievements.

- We celebrate success.

Date:

Trust forms the basis for the relationship charter. Often the guiding principles are developed from the answer to the question: What does trust look like for this relationship? Does the relationship pass the handshake test? In other words, are the parties involved in the relationship prepared to fall back on the trust and principles underpinning the handshake in the event of problems and opportunities arising. It is the relationship charter, as the moral agreement, that will guide your actions and behaviours.

Elements of the relationship charter

There are many variations but typically relationship charters consist of a jointly agreed vision, or mission or purpose statement, a non-prioritised list of key objectives or critical success factors and guiding principles and behaviours.

Richard Whitely, business author, defines a vision as a 'vivid picture of an ambitious, desirable future state that is connected to the customer and better in some way than the current state'.[4] The vision will have wholeness and integrity, and embody the values and long-term aspirations of the relationship. The mission or purpose statement is more specific in timeframe and intent. Usually a single sentence, it outlines what is to be done, the broad goals and who is to achieve them. The main objectives, or critical success factors (CSFs), transform the vision and broad goals into specific aims for the relationship. They are the headline statements for more detailed relationship action plans. The key metrics or key performance indicators (KPIs) are then developed from these CSFs. The principles are those self-evident truths and behaviours that will guide the relationship.

Characteristics of the relationship charter

Ideally the relationship charter is signed by all those people involved in its development. This will certainly include senior management. It becomes the base document by which the relationship performance and improvement is evaluated. The charter also allows commitment to, and ownership of, the implementation and the outcomes and is a genuine symbol of recognition and empowerment for all participants and stakeholders. Table 4.4 provides a summary of the characteristics of relationship charters for each of the 10 relationship types.

Contracts and relationship charters

The relationship charter is not intended to be a legally enforceable contractual agreement. Its purpose is to focus on the working relationship and intent between the relationship parties, not the legal relationship. It is an observation that the more mature and high performing the relationship, the less reliance there is on legal rights and obligations, and the greater the focus on people, trust and win–win outcomes. The legal contract or agreement then becomes more of a safety net or fallback in the event of the relationship deteriorating or collapsing. These legal contracts are seen as bottom drawer contracts, taken out only when issues are to be resolved or clarified.

Table 4.4: characteristics of relationship charters across the 10 relationship types

Relationship type	Characteristics
1 Combative	Relationship charters are not appropriate and not encouraged; they are typically too open and transparent for Combative relationships
2 Tribal	Used to reinforce tribal or group loyalty; mainly internal application; based on what's in it for me (WIIFM)
3 Trading	More to do with company slogans or statements of customer service than charters; buy and sell focus
4 Transactional	More to do with generic, broad market statements of values, and service level commitments
5 Basic	Often implied or inferred or assumed by return or repeat business, that is, integrity based, not typically documented; 'do what you say' focus
6 Major	Results driven and customer centric, with a focus on improvement of cost, quality, schedule, service levels
7 Key	Moral agreement to support legal contract; strategic in intent; not legally binding; a performance management tool
8 Partnering and Alliancing	Strategic engagement document linking purpose, performance, measurement, remuneration and attitudes
9 Pioneering	Framework for the vision or purpose; used to simplify complexity; provides clarity, direction and rules of engagement for the breakthrough relationship
10 Community	Public documents for communication, engagement, public debate and review; a moral agreement with broad community support

Signing these documents is usually followed by considerable publicity, such as media releases, and announcements in company magazines and newsletters, and other internal publications. In raising expectations, the promises implicit and explicit in the relationship charter need to be delivered on. There are consequences for all parties involved should requirements not be met.

The intent of the legal contract and the relationship charter should be consistent and aligned. Care should be taken in switching the focus of behaviour and practice between the two for self-interest or convenience. Doing so to assert authority for short-term gain or to seek one-sided advantage will make the relationship unsustainable.

It is important for legal advisers, who may be key stakeholders, to understand the different characteristics associated with different relationship types. In some relationships, there will be little or no legal involvement. In other relationships, where corporate policy dictates or where size, complexity, dollar value, risks and benefits are significant, the relationship will benefit significantly from the input of legal expertise into both the legally binding contracts and the typically non-binding relationship charters.

The relationship charter is directly linked to the key performance indicator (KPI) metrics, discussed in the following section.

Performance measurement and relationship KPI scorecards

A senior manager once told me, 'In God we trust, everyone else needs data'. That's a good line for a hard-nosed, numbers man wanting to challenge the use of intuitive judgement, but it's not that simple. Data and trust are interconnected. In the hands of the right people, open, relevant, timely and accurate data will create greater levels of trust, and trust is the basis upon which all high performance relationships are built.

Without measuring performance, especially against baselines, it is difficult to know how you are travelling on the journey of improvement. While the old saying 'you can't manage what you can't measure' is self-evident, the questions remain: What should be measured, when, why, how and by whom? This will depend on the relationship approach and the current and future operating environments.

The relationship KPI scorecard is an effective framework for linking relationship goals and objectives to performance (targets) and measurement (results). A simple worked example explaining the connection between the relationship charter and the KPI scorecard is shown through a comparison of figures 4.2 and 4.3. Figure 4.2 shows a sample relationship charter, comprising a vision statement, key objectives and guiding principles for the relationship.

The columns in the performance scorecard (see figure 4.3 on page 136) represent the relationship Key Objectives identified in the relationship charter. The rows represent the six Key Results Areas (KRAs) that are part of the 0 to 10RM Matrix (see figure 1.4 on page 10). In summary, the six KRAs are financial success, customer/stakeholder satisfaction, sustainable competitive advantage, best practice implementation, innovation and attitude. This scorecard framework is then populated with selected hard and soft, financial and non-financial, leading and lagging KPIs and associated KPI targets. In the absence of a relationship charter, key objectives will be found in agreements, business plans and other documents.

The KPIs shown on the scorecard are directly connected to both the relationship key objectives (columns) and individual key results areas, or KRAs (rows). For effective performance measurement, it is necessary to have the right balance between hard and soft, financial and non-financial, leading and lagging KPIs. Not every cell within the KPI scorecard needs to be filled. As a minimum, each column and each row should have at least one KPI to ensure the right representative balance. Getting the right mix of hard, soft, leading and lagging performance measures to capture the current and future health and wellbeing of the relationship is essential. A good KPI scorecard provides an effective measurement tool for evaluating performance in the current state of the relationship and an effective management tool for driving future improvement. As indicated in the key components image in figure 4.1 on page 118, for large and complex relationships there may be sub-level KPI scorecards that roll up into a top level KPI scorecard. For completeness, and in reference to the section below on risk–reward performance measurement, the KPIs and associated targets that are directly linked to supplier remuneration are shown in bold text.

Figure 4.2: sample relationship charter

Relationship charter

Vision

Role model partners making a difference by delivering best practice outcomes and valued solutions through collaborative teamwork.

Key objectives

1 Ensure zero harm to people, plant and environment.
2 Deliver products and services that meet or exceed agreed customer requirements.
3 Achieve sound commercial outcomes for the relationship parties.
4 Innovate for continuous and breakthrough improvement.
5 Build and maintain honest and open communication and information sharing.

Guiding principles

- Act in a way that is best for the business—safely, legally and logically.
- Do what you say.
- Cause no unpleasant surprises.
- Ensure our behaviours are based on trust, integrity, honesty and professionalism.
- Engage proactive problem solving and joint resolution of issues.
- Be fair and reasonable.
- Remain committed to a no blame, high performance culture.
- Have fun and celebrate success.

Figure 4.3: KPI scorecard for the relationship charter shown in figure 4.2

Vision: Role model partners making a difference by delivering best practice outcomes and valued solutions through collaborative teamwork.

Outcomes (leading and lagging)	Key objectives				
	1. Safety health and environment (SH&E)	2. Meet or exceed agreed customer requirements	3. Sound commercial outcomes	4. Continuous improvement and innovation	5. Communication and information sharing
Financial success (ROI)			Customer profit percentage (100% of target) Relationship budget costs (100%)	Innovation dollars achieved as percentage forecast (80%) Innovation budget ($x)	
Customer or stakeholder satisfaction	Total LTIFR (1) Relationship LTIFR (0) Environmental breaches (0)	Service levels IFOTA1 (95%)			Percentage systems performance or access (99.5%)
Sustainable competitive advantage			Percentage available client business (>90%)	Time to market / cycle time (20% reduction)	
Best practice implementation	SH&E plan(s) progress (100%)	Equipment reliability or availability (98%) Project performance (90%)		Number of best practices shared (x)	Actions - overdue (0%)
Innovation				Improvements - identified (x) - implemented (y)	
Attitude	Near miss learnings Implemented (100%) SH&E safety audit plan - actions overdue (0)				Relationship survey results (95%) Unnecessary issue escalations (0/0) Joint training days (x/y)

Notes: KPI (target) e.g. LTIFR (1), Reliability (98%); LTIFR = lost time injury frequency rate

The varieties of KPIs

Hard KPIs are generally the performance indicators that are based purely on objective, quantitative measurement, such as financial measures like return on assets, profitability, operating efficiencies, field tests and other physical measures. Soft KPIs are based more on subjective, qualitative assessment, such as views and opinions associated with community-based KPIs, employee engagement, and surveys of the health of the relationship.

Leading and lagging indicators, through a process of cause and effect, identify the performance output or outcome measures (lagging KPIs) and the performance drivers (leading KPIs). For example, in high performance relationships, lead indicators include early and effective employee engagement and empowerment; expanded information sharing; systems and process integration; implementing near-miss learnings on safety; the number of site visits; joint training sessions; the degree of co-location of partners; and early supplier involvement in the design and planning phases. They lead to lagging indicators such as actual operational excellence, safety performance, reliability, availability, profitability, growth, market share and competitive advantage.

KPIs, like objectives, can be focused on inputs, outputs or outcomes. Take the example of planned maintenance work that is completed on or before time (input KPI) to achieve improved plant or process availability or reliability (output KPI) to achieve the ultimate goal of increased plant yield or factory output (outcome KPI).

KPIs also occur at three levels: strategic, relationship and compliance. Strategic KPIs reflect the overall performance of the relationship and are typically focused on outcomes, such as yields, overall profitability or safety performance, market share and return on investment. They will often involve additional activities outside the relationship scope and control, and so they are more associated with the Type 7 Key to Type 10 Community relationships on the 0 to 10RM scale. Relationship KPIs reflect the measures of performance the relationship is directly responsible for delivering and are often outputs related. They can be in the form of contracted KPIs embedded in binding agreements, or non-contracted KPIs that are part of the broader performance and relationship management process, such as quality, cost, schedule, service levels, functionality, reliability, availability, time to market and innovation benefits. Compliance KPIs represent the lower level, day-to-day measures that are often driven by corporate policy, mandatory requirements, regulatory bodies (for example, environmental protection agencies) and other third parties. They are often inputs focused.

The right set of KPIs is in place when the KPIs start to drive the right attitudes, behaviours and practices among the people who are responsible for delivering the associated business goals and objectives.

KPI scorecards come in different forms and the scorecard shown in figure 4.3 on page 136 can easily be converted to a more traditional table format once the right balance of lead and lag, hard and soft KPIs is arrived at (see table 4.6 on page 144). As indicated in figure 4.1 on page 118, sub-level KPI performance scorecards

---

A coal mining KPI story

One of the best sets of performance indicators I have seen involved the relationship between a coal mining company and the supplier of underground mining equipment. This was a relationship of high impact dollar value. In a worse case scenario, a catastrophic equipment failure could potentially see the mining equipment permanently buried at huge cost.

At the company, the primary measure of performance, which was directly linked to remuneration, was 'feet advanced in the coal seam per eight hours'. This indicator — associated with other KPIs of reliability, availability of critical equipment and machinery, costs per tonne of coal and safety performance — measured clearly and simply what the relationship was trying to achieve in delivering lowest cost coal to the market place.

To deliver to the agreed targets for 'feet advanced in the coal seam per eight hours' required an enormous amount of cooperation, information sharing and innovation between many players. For example, geologists, mining engineers, design engineers, management, maintenance technicians and schedulers, effectively trained and competent workers (also union members) who operated the underground equipment worked in eight hour shifts. Everyone was linked into, driven by and focused on this prime set of KPIs, and everyone was in part remunerated on the achievement of the KPIs. The benefits of having a shared vision, simple and clear common goals and jointly agreed upon accountable performance indicators cannot be underestimated.

Traditionally, the business for the supplier was about selling as much equipment as they could at the highest price and extracting as much from the maintenance and service costs as possible. For the customer, business was about the reverse of that: minimum equipment purchases at the lowest procurement and maintenance costs. Not surprisingly, such conflicting objectives and performance drivers had resulted in poor behaviours and bad attitudes, and created many operational, commercial, communication and interpersonal problems.

may well apply for sub-agreements, additional business cases, projects and internal relationships.

There are different perspectives on performance management and measurement as you move across each of the 10 relationship approaches (Combative to Community). These are summarised in table 4.5.

Table 4.5: variation of attitudes to KPIs and performance measurement across the 10 relationship types

Relationship type	Relationship KPI scorecards and performance measurement characteristics
1 Combative	Win–lose KPIs; short-term profit or financial focus; tough and aggressive; often one-sided
2 Tribal	Defensive KPIs for protecting position and self-interest; used for leverage, assigning blame or finger pointing
3 Trading	Simple cost or price, delivery schedule and quality KPIs
4 Transactional	Simple service level KPIs focused on what, who and when being trackable or traceable; focus on efficiency, speed, process and systems
5 Basic	Focus on IFOTA1 specification input. KPIs predominantly based on quality, cost, schedule, safety; formal and informal reviews of product and service delivery
6 Major	Output KPIs focused on short-term to medium-term cost, quality, schedule, service levels and performance trends against baselines
7 Key	KPI focus on medium-term to long-term, strategic financial and non-financial outputs and outcomes, innovation and process improvement
8 Partnering and Alliancing	Shared strategic KPI scorecard approach, including a focus on legacy, financial and non-financial, interdependent, and leading and lagging KPIs. There is a direct link to a relationship charter and risk–reward sharing, and performance-based remuneration models. Partners ask the question: What three KPIs do we have in common that can be the basis for working together towards common goals for mutual benefit?
9 Pioneering	A few simple KPIs frame the outcomes; a focus on stretch, breakthrough and legacy, linked primarily to improvement and innovation
10 Community	Financial and non-financial KPIs representing community interests and legacy outcomes rather than vested interest goals and objectives

The association between the KPI performance scorecard and risk–reward sharing and remuneration is discussed in the next section.

The scorecard concept originated in the landmark work of Robert Kaplan and David P Norton, recognising business authorities, authors and academics. A full understanding of strategy and measurement would not be complete without a review of their work, particularly their book *The balanced scorecard*.[5]

Avoid conflicting KPIs

There is nothing like a set of conflicting KPIs to quickly undermine performance improvement and dampen morale. For example, in their effort to reduce costs, a major airline some years ago put an incentivised bonus scheme in place for their pilots around the KPI of reducing the fuel burn rate on their aircraft. The less fuel the pilots used below agreed targets, the bigger the bonus. Pilots, being smart people, flew their planes more slowly and minimised the use of air conditioning when aircraft were on the ground and in flight. Not surprisingly, this had a counter effect on other KPIs, such as the percentage of on-time arrivals, customer satisfaction levels and baggage that missed connections. These KPIs were not the responsibility of the pilots — rather, customer service and baggage-handling departments were responsible. A case of well-intended, fair-minded and reasonable people misaligned in their targeted metrics. The conflicting KPIs caused a great many problems and lost revenue before the problem was solved by instituting a one-team approach and aligning KPIs, and putting the focus back on the customer.

At a more general level, the supplier KPI of profitable growth with increased margin may well be in conflict with the customer's metrics and desire to reduce volume offtake or to seek alternative supplies at lower prices. Metrics such as these are not necessarily mutually exclusive, but they do require alignment and a high degree of openness, honesty and transparency between the parties of a relationship. That is, they require a high degree of trust.

Risk–reward performance-based remuneration models

Learning to love your partner's profit is central to the effective sharing of risk and reward, and sound performance-based remuneration. This is not about begrudgingly accepting, condoning or tolerating the other party's right to profitability but positively and directly linking their success with your success. Better than expected performance delivers better than expected financial and non-financial outcomes for all parties. For many, the outcome of lower customer costs and higher supplier profitability, or better product or service performance at a lower cost is counterintuitive. The reality is that, in high performance relationships, where there are high levels of trust and transparency, the more money the supplier makes, the better off the customer will be. Performance-based remuneration is not a panacea, but when it is well implemented and for the right reasons it is an effective tool in aligning goals, attitudes, mindsets, behaviours and performance outcomes.

Before adopting risk–reward based remuneration, it is essential for the right people to be in place, with the right attitudes, doing the right things in the right way — in effect, Principle 4 is at work. The sharing of risk and reward and linking return to performance adds the focus and the incentive to maximise mutual benefit from a shared purpose and a common set of goals.

Performance-based commercial frameworks can apply to all relationship types from Type 1 Combative to Type 10 Community. When you think about it, all customers and suppliers have their profit at risk in one way or another, be it based on fixed price lump sum arrangements, unit price, schedule of rates or open-book performance based. Profit can be at risk based on downside penalties (for example, liquidated damages, or additional costs associated with re-work, defects or time delays) or upside incentives (for example, cost savings, process improvements and innovation). Performance is critical to all of these. The real issue is not profit at risk but the level of openness and transparency that exists. This is based on competence and the character of the relationship participants around which the risk and reward are shared, and the details are communicated and managed.

I have observed many successful Type 8 Partnering and Alliancing relationships and projects based on fixed price, lump sum arrangements, or other non-gainsharing, non-painsharing commercial arrangements. In these cases, the early engagement, open and transparent approach, mindset and attitudes of the people involved, as well as aligned business drivers and relationship value propositions, were all crucial factors to success.

That said, performance-based remuneration is one of the main mechanisms for ensuring that the value propositions are delivered, and the business drivers and objectives of the relationship parties are aligned. Having skin in the game provides an essential link at all levels between performance, measurement, behaviour or attitude, risk management and remuneration. This creates a focused, one-team culture and an environment of open communication, transparency and a proactive willingness to deliver agreed common goals for mutual benefit. A 'them and us' responsibility is turned into a 'we' accountability.

So how does it work? Figure 4.4 (overleaf) provides a simple image of how return for effort (gain or loss) is linked to performance.

Typically, the supplier's financial remuneration and financial success are linked directly to the customer's own success in the market place or the performance or effect of the products and services delivered. The supplier is, therefore, rewarded on the quality of the solutions and benefits generated, not on the cost or features of the purchased technology, products or services. That is, the supplier earns profit on the quality of outputs and outcomes, not on the cost of inputs. The percentage of agreed profit or revenue, gain or loss, above and below the agreed target, is based on the over-performance or under-performance against agreed KPIs. The extent to which the profit or revenue will be put at risk will depend on the scope of the business case or project activities, and the negotiated outcome of the risk profile both the customer and supplier are prepared to accept. One of the most common variants of this model is the direct cost reimbursable model. In this model, direct costs, such as direct labour, salary burdens, mobilisation and demobilisation costs, direct overheads, selected third party costs and materials, are not at risk and are reimbursable to the supplier.

Figure 4.4: the link between return on effort and performance

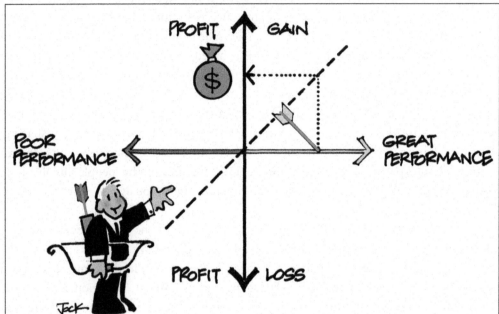

Other commercial elements, such as indirect costs and corporate overheads, profit and reward components, are potentially at risk on a percentage basis as negotiated and mutually agreed between the parties. Figure 4.5 is a simple diagram of this model.

Gainsharing and painsharing, and risk–reward performance-based remuneration is a basic element in demonstrating ongoing value for money in a relationship. This open-book sharing should minimise or put to rest any concerns around uncompetitive and non-contestable pricing, secretive back room deals, or collusive or anti-competitive behaviour. Joint benchmarking, and transparency of appropriate and agreed market information provide an important element of competitive assurance demonstrating a new paradigm approach to contestability. That is, the contest is no longer between customer and supplier, but rather their relationship and the market place.

Table 4.6 on page 144 shows a simple risk–reward based performance scorecard. It is a continuation of the worked example of a scorecard shown in figure 4.3 (on page 136). The strategic and relationship risk–reward based KPIs previously identified are shown on the left-hand column. The next column on the table shows the percentage weightings (totalling 100 per cent) associated with each of the KPIs. Next are the lower, target and upper performance levels and profit levels set for each KPI. This is followed by the actual score for each KPI and the calculated profit level. The totals are then shown at the bottom of the table, along with the total relationship performance on a 0 to 200 per cent scale. This ongoing measurement process is used to track and manage performance over time and, therefore, plays a key role in the continuous improvement process.

Figure 4.5: direct cost reimbursable model of risk–reward performance remuneration

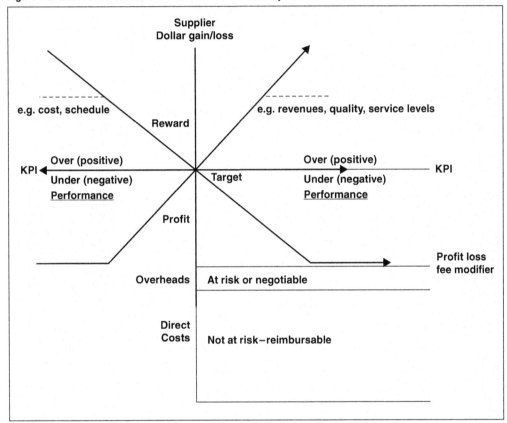

The quantitative metrics shown in table 4.6 (overleaf) can then be used as the basis for plotting relationship performance on the vertical axis scale of the 0 to 10RM Matrix (figure 1.4 on page 10). Figure 4.6 on page 145 shows the overall result in graphical terms.

Table 4.6: a KPI risk–reward based performance scorecard

		KPI performance levels			Profit targets			Actual score	Actual profit
	Weighting	Lower	Target	Upper	Lower	Target	Upper		
Strategic KPIs									
Customer profit as a percentage of budget	10%	0%	100%	200%	0%	0.8%	1.6%	140%	1.1%
Total safety LTIFR	15%	2	1	0	0%	1.2%	2.4%	1	1.2%
Relationship KPIs									
Budget direct costs	20%	120%	100%	80%	0%	1.6%	3.2%	90%	2.4%
Service levels	20%	90%	95%	100%	0%	1.6%	3.2%	97%	2.2%
Relationship survey	10%	90%	95%	100%	0%	0.8%	1.6%	97%	1.1%
Project performance	10%	80%	90%	100%	0%	0.8%	1.6%	85%	0.5%
Innovation and improvement	15%	60%	80%	100%	0%	1.2%	2.4%	90%	1.8%
Total supplier partner profit	100%				0%	8.0%	16.0%		10.3%

Total relationship performance score (0 to 200%)				
Unsustainable	Poor	Good	Excellent	Outstanding
0–70%	70–100%	100–130%	130–170%	170–200%
		129%		

Total relationship performance

Note: LTIFR means lost time injury frequency rate

Figure 4.6: KPI performance graph

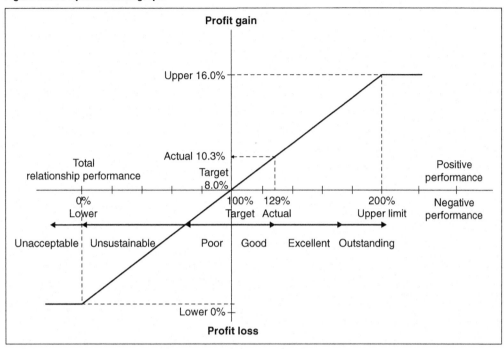

The gain and the pain associated with sharing risk and reward are not confined to supplier profit or margin. Gain and pain can also be associated with repeat or return business, or the term of agreement which can to be extended or reduced based on performance outcomes. Percentage of future available business for the supplier can be increased or decreased in a similar manner. New projects can be bundled into larger projects or work programs, or negotiated rather than taken to the market for tender, again based on performance above or below target. Good organisations are limited only by their imagination when it comes to how they share benefits.

Getting paid on performance or effect

If you are in the business of solution selling and you are good at it, you should aim to be paid on the basis of the solution not the cost of delivering it. If you're going to focus on total value and on improving the margin, why not get paid on the same basis? The really clever organisations are paid not on the basis of unit price but on value — that is, the performance or effect of the products and services delivered. Some examples would include selling aircraft tyres on the basis of dollars per landing, not dollars per tyre; energy companies being paid as a percentage of the customer's energy savings, not dollars per hour of engineer time to achieve those savings; a maintenance company being paid for every hour that plant and equipment is running, not on a dollars per hour basis to get it running when it breaks down; or the same maintenance company getting paid a percentage of plant hourly output. Other examples could include doctors who are paid for keeping people healthy rather than getting them healthy; and lawyers earning a percentage of winning outcomes rather than an hourly or daily fee. It's all about getting paid on the basis of performance and effect, not unit price. The simple principle of getting paid on uptime instead of downtime still eludes many maintenance and manufacturing companies.

The explosives company that doesn't get paid on the basis of dollars per tonne of explosives delivered to the mine site (that is, unit cost) but on dollars per tonne of broken rock delivered to the treatment plant is another example. Effectively selling a specific particle size distribution of blasted rock within agreed environmental limits demands more open communication and information sharing with all stakeholders and makes for a more efficient operation and higher output of the treatment plants. The result is improved supplier profitability at reduced overall production costs for the mine operations. That's a genuine win–win outcome linking all three aspects of customer intimacy, operational excellence and product innovation. The examples are almost endless.

There may be many points of differentiation beyond the direct product or service delivered. Type 8 Partnering and Alliancing relationships in particular involve a long-term package of products and services plus the other tangible and intangible benefits that make up a relationship. These other benefits may involve areas that are strategic, material or personal, and they may have no direct bearing on the immediate delivery of the product or service. Combined, however, the suite of benefits may have an enormous influence on the customer's short-, medium- and long-term buying decisions and, ultimately, on the quality of the relationship itself. Quality, industrial relations, IT solutions, staff learning and development are just some areas for consideration.

The connection between risk and reward, performance and remuneration is far broader in interpretation, application and value than it first appears. There are applications right across the 0 to 10RM spectrum that apply internally and externally to the organisation. Table 4.7 provides a summary.

Table 4.7: characteristics of risk–reward performance-based remuneration models across the 10 relationship types

Relationship type	Characteristics
1 Combative	Focus on risk-transfer, and fixed price or fixed fees or rates; non-transparent; focus on financial aspects and short-term profit; one-sided
2 Tribal	Not transparent; risk averse or risk linked to fixed price, or fees, or rates schedule
3 Trading	Risk and reward separated and not shared; more downside penalties than upside rewards
4 Transactional	Risk–reward performance-based remuneration models are the exception rather than the rule; pay and use focus; standard terms and conditions cover risk; fixed price or fee includes a reward or margin
5 Basic	Typically performance-based remuneration is in the form of return or repeat business based on performance delivered and competitive, profitable costs or fees
6 Major	Linked to KPI performance against benchmarks; focus on direct costs and total cost of ownership
7 Key	Direct risk and reward are managed and measured at tactical or operational and strategic levels; focus on innovation and solutions
8 Partnering and Alliancing	Open, transparent, shared risk and reward based on gainsharing and painsharing; over or under performance achieved against agreed KPIs; operational, strategic and legacy based; remuneration is linked directly to the performance and effect of the products and services delivered by the supplier or service provider
9 Pioneering	Risk–reward inherent in the breakthroughs required; Pioneering parties often bet their future on faith, passion and confidence around unclear risk profiles
10 Community	Open, transparent, shared risk and reward; individual responsibility, collective accountability, community benefit

Relationship plans

It's difficult to see the big picture when you are sitting inside the frame. To reach the desired future state for high performance relationships will require a plan, irrespective of the relationship type involved. The complexity cannot be managed on the back of a postage stamp. The relationship journey management plan is the central document in linking vision, strategy, actions, people and priorities. This plan is a living document and roadmap that will change over time according to the nature, development and performance of the relationship. Thoroughness in the planning stage will be rewarded. A summary of how these plans present themselves across the 0 to 10RM relationship types and their characteristics is shown in table 4.8 (overleaf).

Table 4.8: characteristics of relationship strategy business or action plans across the 10 relationship types

Relationship type	Characteristics
1 Combative	Each party in the relationship has separate strategies or action plans; tightly managed and controlled by each party
2 Tribal	Each party in the relationship has separate strategies or action plans; demarcations established; secretive and controlled
3 Trading	Strategies or simple action plans cover what, who and when; trackable, traceable, deal based; focus on efficiency and effectiveness
4 Transactional	Systems, process and technology driven; often integrated into market segments, sectors, business functions; depersonalised
5 Basic	Focus on simple what, who and when action plans
6 Major	Linked to KPI measures, and contractual obligations and supply chain analysis
7 Key	Strategic plan leads relationship development and performance
8 Partnering and Alliancing	Joint business plan; flexible and adaptive; based on performance and improvement; involving integrated processes; linked to Partnering and Alliance relationship charter
9 Pioneering	Joint business or action plans built on the sense of urgency and breakthroughs required
10 Community	Business plan based on the community; triple bottom line and legacy focus; common purpose; common goals set for the common good

Table 4.8 gives an overview of the different approaches to documentation around relationship plans. Two sample relationship plans follow: a detailed Type 7 Key relationship plan for a customer profile (starting overleaf), and a joint relationship business plan for a Type 8 Partnering and Alliancing relationship (pages 153 to 154). The plan content outline can be adapted and modified as appropriate to the time, the products or services under consideration and the current market conditions. For high performing Key relationships, the relationship plan will be a shared and jointly agreed document. Key relationships are associated with a *do and add value* approach, based on agreed and aligned strategies and exploitation of synergies. Some areas of information or data may be withheld by each party to the plan for commercial confidentiality reasons or what is regarded as company sensitive information. While typically initiated and developed by the supplier party in the relationship, the Key relationship plan can be initiated and developed separately or jointly by the relationship parties. The essential elements of a comprehensive Type 7 Key relationship plan would include:

- An executive summary of the relationship outlining the business strategies of each party in the relationship and detailing the underpinning value propositions

- Corporate or organisational structure(s)
    - organisation chart(s), roles and responsibilities
    - who are the decision makers, key stakeholders, key influencers and other key players?
    - leadership style(s)

- Financial performance

- Relationship governance

- Relationship charter

- Current business and future opportunities, plus SWOT analysis

- Customer or supplier business in detail (for example, market segment strategy, products, services, facilities)

- An answer to the value question: What value is this relationship delivering for our organisation over the alternatives?

- Specific action plans with performance metrics over the next 12 months, two or three years.

The style, content and form of the action plan will vary depending on the nature of the relationship, considering such variables as market sector, customer, supplier and competitor, or whether the focus is internal or external. For large global Partnering and Alliancing relationships, these plans may involve detailed business cases and major project plans.

Key relationship plan — customer profile

Prepared by:
Date:
Plan period:
Customer name:
Supplier name:
Main contact details:
e.g. physical and postal address, phone, fax, key email addresses
Website(s):

Plan executive summary, purpose and intent
Business overview or background to the relationship, including such information as the age of the relationship (in years), an overview

(continued)

Key relationship plan — customer profile *(cont'd)*

of importance of the relationship to the organisations (that is, strategic value and commercial value) and the underpinning relationship value propositions. Outline of the general business strategy for the relationship parties, such as expected growth rates, market sector focus, and current and future operating environment.

Organisation structure
Consider the following points:

- What is the ownership, company type, organisation (shown through company charts), internal value chain or primary and support activities process flows?

- Who are the decision makers, and what are their roles, responsibilities and personal interests?

- Who are the key influencers, key stakeholders, champions, mentors and coaches?

- What are the business style(s) and leadership style(s) operating?

- What are the relationship maps between and within the relationship parties?

Financial performance
- Consider profitability, return on investment (ROI) and other financial ratio analysis data, the latest balance sheet and profit and loss data, and annual report information.

- Business done with other parts of the organisation (if appropriate).

Reciprocal business:
- The nature of any two-way buying and selling of products and services that may exist between the relationship parties.

- Volumes, value, any special terms and conditions, and so on.

Relationship governance
- Overview of how the relationship is managed and governed over time (that is, the details surrounding stewardship for the relationship and the reporting mechanisms).

- Relationship teams or work groups—membership, roles and responsibilities, meeting frequency, including lead team, management team, wider integrated project teams.

- Business review and development meetings (frequency, such as six monthly, and agenda items).

Relationship charter
Vision and purpose, key objectives, guiding principles
Current business and potential opportunities associated with relationship objectives, including:

- product or service supplied—names or listings

- products or services produced

- sales history (volume and dollar value), including historical data and trend lines, current state and forecast or budgeted volumes, dollar values

- current prices, credit terms and conditions, rebates, incentives operating

- current percentage of available business (volume or dollar value): who has the balance of the business and what are the products or services provided (volume or dollar splits)

- projected market growth rates

- competitors (products, prices, volumes, application or effect) their performance rating as a supplier and why they are so rated

- SWOT analysis (strengths, weaknesses, opportunities and threats to your organisation, customer organisation, competitors)

- customer current buying criteria or requirements.

Customer or supplier business in detail
Consider the following areas of the business.

Strategy
- vision or mission statement and values statement

- corporate objectives and enabling strategies

- division objectives or strategies

(continued)

Key relationship plan—customer profile *(cont'd)*

- product group objectives or strategies

- key issues (internal and external)

- development plans or opportunities and your current involvement

- competitive advantage or point of difference

- SWOT analysis.

Marketing and sales
- downstream products or services

- their application or effect

- sales volume or dollar turnover

- the market sectors served and market share

- a picture of the external supply chain, upstream and downstream

- growth rates

- major competitors (for example, names and market sectors, market share).

Manufacturing or service infrastructure (by site)
- relationship maps showing the nature of the internal networks that exist within and between other functions and departments

- plant and equipment (for example, type, capacity, age, condition)

- process description

- support facilities (for example, training and development, technical service, legal, financial, commercial research and development).

Answer the value question
For the customer and supplier: What value is this relationship delivering for our organisation over the alternatives?

Results of relationship health check analysis
Conduct the check using the eRAD or RAD (see chapter 2).

Action plan

This should consider the next 12 months, two years and three years in the following terms.

Focus areas, objectives and critical success factors

- What has to be done? Who is the owner? What is the completion date?

- What are the resource requirements?

- What are the associated KPIs?

- What is the current progress or performance against the plan?

Business review and development meetings

Consider dates, participants, minutes of actions and status, and the performance scorecard of current performance.

Joint relationship business plan for a Partnering and Alliancing relationship

A joint relationship business plan would typically have a three-year timeframe, with one-year horizons as interim milestone relationship evaluation points, or alternatively conduct regular half-yearly reviews. It should include the following contents:

- Overview or background to the relationship business plan or the journey management plan.

- Operating environment — both individual and joint.

- The relationship charter and its vision, key objectives and guiding principles.

- Governance and team structure, including people, and their roles and responsibilities, and the decisions they make.

- Relationship performance and progress to date and RAD or eRAD review (0 to 10 relationship alignment diagnostic).

- SWOT analysis and establish the critical success factors (CSFs).

(continued)

Joint relationship business plan for a Partnering and Alliancing relationship *(cont'd)*

- Strategies, initiatives and activities over one, two and three years, shown in a table; noting what is to be achieved, who (is the owner), and by when to be completed.

- Measures of success — KPIs or targets or desired outcomes (based on the KPI scorecard); financial; customer or stakeholder satisfaction; sustainable competitive advantage; best practices; innovation; and attitude.

- Answer to the value question: What value is this relationship delivering for both organisations over the alternatives?

- Other areas, including resources; sub-level business cases and projects; financial or capital analysis; risk management analysis and mitigation; benchmarking; stakeholder management; specific culture, strategy, structure, process, people issues and opportunities; communication, multi-level interfaces, contact details; recognition and celebration of success.

Other aspects to consider around relationship plans

- What is the target audience(s) for the plan, such as management, operations, stakeholders?

- What is the timing and makeup of the development, sign-off and review process?

- How will the plan be communicated and implemented?

The importance of relationship plans

Relationship strategy or action plans are crucial for a number of reasons:

- Plans provide a roadmap for relationship management and improvement as well as providing big picture documents linking strategy and tactics across the short-, medium- and long-term.

- Plans have wide visibility in the organisation(s), bringing together the cross-functional, cross-organisational, interdepartmental activities that affect the relationship. Having plans in place can remove roadblocks and help to break down departmental barriers if they are used effectively. This enables early alignment and buy-in from key personnel, key influencers and relationship champions.

- Plans form the central document for the relationship manager(s) in coordinating activities and managing the progress and performance of the relationship.

- Plans provide a database facility and corporate memory for the relationship as people move in and out of the relationship.

Performance against the plan is a direct measure of the quality of communication and ownership of the relationship. For example, tracking the number of outstanding actions and overdue actions as a percentage of total actions is indicative of whether people are doing what they said they would do. Early warning and extensions of deadlines may be given, but too many extensions indicate a lack of commitment to getting things done, and the warning flag should be raised. I remember a high impact relationship in the early stages of development where the target for overdue actions was less than 5 per cent but the actual figure was 55 per cent. This was, of course, generating angst and frustration. At an ensuing business review and development meeting, corrective actions were successfully implemented.

See chapter 7 for further detail about the relationship strategy and action plans, in the section on the 12/12/6 roadmap steps.

Governance and structure

Governance is the process of stewarding, leading and managing the relationship for improvement. A reminder about the connection between strategy and structure: it is the business and relationship strategy that should determine the structure of the relationship, and not the reverse. Governance structures and associated people responsibilities will take on different forms depending on the relationship type engaged, and the size and complexity of the organisations involved. These could range from single points of contact or automated electronic contact in vendor relationships, to dedicated account management and multi-level contact structures in supplier segment relationships; and to team-based and multi-leadership and management structures in the partner segment. Relationship structure and relationship interfaces as they apply to the vendor, supplier and partner segments are illustrated in figure 4.7 (overleaf).

Figure 4.7: relationship interfaces and governance across vendor, supplier and partner segments

Governance structures and characteristics across the 0 to 10RM relationship spectrum are outlined in table 4.9.

Table 4.9: characteristics of governance and structure across the 10 relationship types

Relationship type	Characteristics
1 Combative	Command and control governance structures; hierarchical; hostile interfaces and abrasive rub points
2 Tribal	Defensive, protective, secretive, territorial interfaces; informal contacts and networks; backroom deals; closed and often non-transparent governance
3 Trading	Simple or single point of contact interfaces; face-to-face or electronic communication; governance often involves dominant deal makers
4 Transactional	Electronic or single point of contact; arms length; impersonal technology-based communication
5 Basic	Simple, single or limited points of contact; work to rule; review is face to face or electronic

6 Major	Regular formal performance reviews at mid senior levels led by the major account manager(s); medium to high level contract management interface or focus
7 Key	Work or project teams directed by key relationship managers and senior sponsors; stewardship from senior managers; regular business review and development meetings, for instance, six monthly
8 Partnering and Alliancing	Joint ownership, accountability; one-team approach; joint leadership and management teams with executive sponsorship; regular relationship performance reviews
9 Pioneering	Pioneers, paradigm shifters tend to be dominant players within small, tight, team-based governance structure(s); special purpose teams often in place, reporting directly to executive sponsors or steering committee
10 Community	Extended supply chain and community governance around a community of practice; democratic community councils; loosely defined or unclear governance structures associated with social community networks; open source style governance structures

Governance structures and associated roles and responsibilities for high performance Type 8 Partnering and Alliancing relationships are discussed further in chapter 7.

Summary

- The 0 to 10RM key components simplify the complexity of relationships.

- The eight key components that need to be managed are:
  - Leadership value propositions
  - Relationship contracts and legally binding agreements
  - Relationship charters as the moral agreement
  - Relationship key performance indicator (KPI) scorecards
  - Risk and reward performance based remuneration
  - Relationship strategy, action and business plans
  - Relationship governance.

- Principle 3, Customers are the reason suppliers exist, supports the application of the key components to manage the details of the relationship.

- Value propositions are those compelling opportunities and benefits beyond a cheap cost and low price. They will arise from the understanding of strategic and commercial value of the products and services associated with the relationship.

- Good leadership combined with compelling value propositions are needed to effectively manage the key components and the associated relationships for better business outcomes.

- You can't manage what you can't measure. However, engaging a new paradigm and embarking on a journey to a new destination may require a leap of faith.

- Trust is critical in any high performing relationship. Understand what trust looks like for each relationship. This insight forms the basis of the guiding principles in the relationship charter and is revealed in people's attitudes, behaviours, practices and performance.

- The effective management of the key components is fundamental to strategy implementation and sustaining competitive advantage.

- High performance relationship management needs to be underpinned by a clear strategy and set of value propositions that give reason to the nature of the relationship type and performance level engagement — the relationship approach is a means to an end, not an end in itself.

- Strategy should determine structure and not the reverse.

Part B

PEOPLE AND CHANGE

CHAPTER 5

LET'S GO CHANGE MODEL

The Let's Go change model (Theme 3) mobilises the Bus of Change and the people in it, allowing the journey of relationship rescue, improvement and transformation to take place. The Let's Go change model (see figure 5.1, overleaf) is a call to action.

I keep six honest serving-men
(They taught me all I knew);
Their names are What and Why and When
And How and Where and Who.

Rudyard Kipling[1]

Managing change is simply the process and journey around bridging the gap between a current state and the desired future state. It implies varying degrees of adjustment, reorganisation, restructure and, often, transformation. The link between change management and 0 to 10 Relationship Management can be seen in the 0 to 10RM Matrix (figure 1.4 on page 10). Managing the 'how to' journey of change between the current state and desired future state is the basis upon which relationships are managed and improved.

People — you can't do business without them

Efficiency is doing things right. Effectiveness is doing the right things.

There is almost nothing worse than doing something efficiently that doesn't need to be done at all. Principle 4, The right people doing the right things, in the right way, at the right time, for the right reasons will deliver the best results, is the rightness principle, where effectiveness and efficiency are constantly being balanced, recalibrated and re-engineered. Running outdated technology efficiently is unlikely to have the leap-frog effect in improving speed to market of new products and services or generating competitive advantage. As the old story goes, there is little point turbo-charging the piston engine when your competitor has just invented the jet turbine.

Figure 5.1: Let's Go change model

As humans, we have an overwhelming desire to communicate. It is part of our DNA. We are hard wired to build relationships. How we communicate and what relationships we form will determine the results we achieve — for better or for worse. The good news is that relationships have little to do with rocket science — that is, technical complexity. Be they business or personal, relationships are all about people.

People come in all shapes and sizes, different likes and dislikes, all varying in their motivations, idiosyncrasies, personalities, ambitions, skills, capabilities and needs. Technology and processes are important but without the people there is often no reason and no means. Having the right people doing the right things, the right way, at the right time, for the right reasons aligns all the elements of relationships, processes and technology. Managing relationships is all about applying common sense. Unfortunately, common sense is not always common practice.

Principle 4 is also directly linked to change and change management. The Bus of Change illustration (see figure 5.1) links the principle to the process and journey of change. Just as there is diversity of people in relationships, people also differ in their ability to cope with and manage change. The Bus of Change infers that there are broadly four categories of people involved in the change process: innovators, early adaptors, followers and saboteurs.

While in no way diminishing the importance of cost and value for money, if Principle 4 is effectively applied, in particular in the early concept, design, negotiation and transition phases of relationship development or change, then there will be a greater likelihood of financial success. Applying Principle 4 will also provide more opportunity for innovation, problem solving, stretch and breakthrough improvement in the execution phases. This will in turn maximise the probability of delivering the best possible outcomes.

The Bus of Change

The Bus of Change is one of the most enduring concepts within 0 to 10RM. It has proven to be helpful, effective and non-threatening in engaging a discussion around the different types of people, attitudes and behaviours that define relationships.

It is not the strongest of the species that survive, nor the most intelligent, but the one most adaptable to change.

Charles Darwin[2]

In his book *Relationship marketing*, Regis McKenna states that 'social science researchers have noted that people can be divided into four categories according to how quickly they adopt new products and beliefs'[3] — that is, how quickly they adapt to change. The four categories are innovators of change, early adoptors to the change, late adoptors and laggards. McKenna says: 'According to one book on the subject, about 2.5% of the public are innovators, 13.5% are early adopters, and 16% are laggards'. I have adapted and modified this language into the Bus of Change concept and the Let's Go change model. I have substituted early adaptors for early adopters, followers for late adopters, and saboteurs for laggards. Followers, at 68 per cent, are the majority and they are doing exactly that — following.

One of the objectives of all good businesses and organisations must be to encourage and lead the change profile towards the innovators. Adaptability to change will be one of the essential factors in achieving effective high performance relationship management.

Encouraged and empowered, innovators will provide the activation energy and ideas to overcome many of the hurdles and explore many of the opportunities. They are the change champions, provocateurs and leaders for the relationship improvement process; they will not only imagine the future but also play a critical part in creating it. They have strong intuitive judgment. That is, they have the ability to make good decisions based on little or no information — they often work on the basis of 'it just feels right!'

Early adaptors keep the momentum going. They are usually the first group of non-innovators to commit themselves to the change process. They are not above questioning, challenging, modifying and improving the change as required in order to accommodate the greater majority of people into the process more effectively. In this way early adaptors genuinely add value to the innovation or change process. However, they do not automatically adopt whatever it is that the innovators want to implement. Often acting as a bridge, they will stop the followers from losing sight of the innovators. The classic early adaptor statement is 'Wow, what a great idea. I wish I had thought of that!' This group needs to be encouraged, coached and led. The earlier they catch on, the quicker the pull through effect and the faster your organisation's progress will be.

Joel Barker talks about 'paradigm pioneers',[4] a special group of people who drive the paradigm shift from rough concept to practical application. Paradigm pioneers have the courage, intuition and long-term perspective to make the vision a reality. They may be found anywhere in the organisation, at any level, and in every function. They can also be third party, trustworthy independents outside the organisation. In fact, Joel Barker suggests the person most likely to change your paradigm is an outsider, as they have little investment in the prevailing paradigms. Paradigm pioneers will be found among the innovators and early adaptors, developing, creating and discovering.

Followers are in the vast majority. They know a good thing when they see it, but their enthusiasm to support change will depend directly on the quality of the argument and the delivery of results. In most organisations they are highly skilled and committed individuals doing a good job, but they are not early adaptors or innovators of change. One of the real tests of how well a relationship is going is to what degree the followers are on side: how engaged, empowered and committed are the followers, and to what extent are they involved and participating in the relationship improvement process? There should be no reason why selected followers and early adaptors can't move up the scale to become early adaptors and innovators respectively.

Saboteurs are different people altogether. This label creatively describes those individuals who overtly, or covertly, oppose or deliberately undermine the change process. They are the gatekeepers, road blockers and filters in the business context; in terms of high performance relationship management, they will ignore the principles and reject the process. At best this presents as passive resistance. Typically there is some degree of deliberate and active resistance. In other words they are continually looking for an opportunity to throw a spanner in the works, fouling things up and sabotaging the efforts of more fair-minded and reasonable people. Saboteurs can reside anywhere inside or outside the organisation and range from senior management to the shopfloor, in any department or any function, from operations to sales and marketing.

Saboteurs often live in an environment characterised by tribalism and mistrust. This can be a breeding ground for power-hungry, self-motivated individualism, self-interest, complacency and arrogance. They can't be overlooked, since they have the potential to stall or impede the progress of even the best performing relationships. That is, they need to be managed up or out, or at the very least be working in a position where they can do little harm. Many, however, will just move on as the change will be too great for them to handle. The first and best strategy for dealing with saboteurs is to engage them in a positive, open manner. Listen to their viewpoints and opinions. Harness their emotion and energy, and identify and agree on common goals for mutual benefit.

The similarity between innovators and saboteurs is that both groups are made up of stubborn, passionate and unreasonable people. It is these qualities or attributes, focused in the right — or wrong — direction, up the front or down the back of the Bus of Change, that allow them to have a constructive or destructive impact on the change process. Don't be fooled, saboteurs are not stupid people. They are, however, often misaligned with the strategy and direction of the change initiative and display the wrong attitudes, behaviours and approach. There is nothing worse for an organisation's performance than smart people coming to work with a bad attitude.

High performance relationship management is a complex mix of human behaviour and organisational, market and technological diversity. This is what makes customer/supplier relationships challenging to develop, hard work to sustain but rewarding and enjoyable when they succeed. People and communications are the building blocks.

Let's Go change model explained

Managing change is important because our lives, the environment in which we live, and the business we work in are in a constant state of evolution. Change is everywhere in our lives, big and small, critical and non-critical, short-term and long-term. Changes can include the challenge and excitement of changing jobs to advance career development; losing weight or giving up smoking for better health; or relocation to another city or country to make a better life. At both a personal and business level, how effectively we manage change, and our attitudes and behaviours will determine both the quality of our lives and the delivery of business outcomes.

How well we understand the need for change, and are informed, engaged, empowered and possess the required resources and capabilities, will ultimately determine the success of any change initiative. Ideally, change is about progress, improvement and a shift to a better place. Change will involve a number of variables, from vision and leadership through to willingness and capability at all levels in the organisation. Change will also depend on each individual's perspective and position in the change process. Change management, like relationship management, when executed well, will lead to a better quality of life, better relationships and better business outcomes.

Some people embrace and accept change willingly, exploring the options and opportunities. Others fear change, seeing it as a potential threat to their world. Denial and resistance are often expressions of this fear. Past experiences will colour the way an individual views and manages change. Having a process to effectively manage the journey of change will alleviate the concerns associated with its implementation, allow testing and exploration and, in turn, foster acceptance and buy-in.

We have all had a friend or business colleague who in a state of excitement has said 'Let's go ...'. In the social, business or organisational context, Let's Go is often followed by a support or scope statement like those set out in table 5.1. For example 'Let's go on a journey. Let's go do something different.' Typically, our immediate response involves one of the Five Ws — Where, Why, What, Who, When. There can also be two other possible responses to the Let's Go phrase. They are Okay and No. Both these responses infer the Five Ws have been dealt with to a satisfactory degree. Table 5.1 looks at a sample of Let's Go statements, expands the Five Ws and their link to Principle 4, The right people doing the right things, in the right way, at the right time, for the right reasons will deliver the best possible outcomes.

Table 5.1: the links between typical Let's Go statements, the Five Ws response questions and Principle 4

Let's Go purpose statements	The Five Ws	Five Ws links to Principle 4
• on a journey • and take a different strategy or approach • and change the way we do things • and re-organise • and re-structure • and do something different	**Where** are we going? (from and to)	To deliver the best possible outcomes
	Why are we going?	For the right reasons
	What do we have to do to get there and **how**?	The right things, the right way
	Who is involved?	The right people
	When do we start and finish and when are the milestones?	At the right time

The challenge is to understand and manage the complexity involved in the subsequent journey of change from the Let's Go starting point, through the transition milestone(s), to the final destination or end point. The Let's Go theme and the Five Ws are a simple and effective tool for managing change initiatives. Before starting out on the Let's Go journey of change at an individual, work group and organisational level, we often need to let go of information, past attitudes, behaviours and practices.

Strategic and tactical Five Ws

In practically applying the Five Ws, there are two levels of engagement: strategic (design and higher level planning) and tactical (execution and delivery). The relationship between the strategic and tactical Five Ws is shown in table 5.2.

Table 5.2: the relationship between table 5.1 and strategic and tactical Five Ws

Strategic Five Ws		Tactical Five Ws
Where		Where
Why	How →	Why
What		What
Who		Who
When		When

Let's Go action plans

The Let's Go 'how to' action plan involves deciding on the strategies, initiatives, actions and timelines by which the gap between the current state and the desired future state will be bridged, and how the secondary states will be effectively managed.

The Let's Go action plan is demonstrated in the case study of Suncor Energy. It is an excellent example of how the Let's Go framework can be implemented practically to ·assist in the delivery of improved business outcomes. Journey to Zero (JTZ) is Suncor Energy's vision for a workplace that is free of all occupational injuries and illnesses — zero injuries.

Suncor Energy is a large integrated energy company strategically focused on developing Canada's oil sands. Suncor's operations include conventional and offshore oil and gas production, petroleum refining and product marketing under the Petro-Canada brand, and a growing renewable energy portfolio. It should come as no surprise that safety is all-important for Suncor, and it cannot be compromised for any reason or at any cost.

The following story was written by Sandra Kendel, senior process improvement specialist at Suncor Energy's In Situ Business Unit, Firebag site. The Suncor team at In Situ have used the Let's Go change model to good effect for better business and safety outcomes. The accompanying strategy on a page for Suncor In Situ 'Let's Go' Journey to Zero is shown in figure 5.2 on page 170.

Suncor case study

Let's Go Journey to Zero for the In Situ Business Unit

You can have the best idea, one that is genuinely unique and innovative, but if you can't effectively communicate that idea, it is destined to remain an unrealised dream. Using the Let's Go change management model, Suncor In Situ has developed a strategy to implement important cultural change relating to safety.

The leadership team saw this change initiative as fundamental to imbedding safety best practice and behaviours into 'the way we do business' at Suncor. Now, a year and a half into the program, it is both sustainable and beneficial to the organisation with employees showing significant interest.

Our In Situ business unit has three components: the Firebag site, the MacKay River site and Resources. It is a complex work environment, challenged by the remoteness of our operations in northern Alberta, the newness of our In Situ technology and a geographically dispersed workforce. Yet, despite these challenges, we are steadily improving our safety record, which ultimately results in superior employee engagement and operational excellence.

Our safety program called 'Journey to Zero' (JTZ) began in 2003, at the Suncor Oil Sands Business Unit at Fort McMurray. At In situ we have built on the work done at our Oil Sands business unit and are now in a position to deliver world-class safety outcomes. Using the Let's Go change model, we were able to wrap our arms around the different and complex aspects of this culture change and adapt and modify the program materials to be able to facilitate numerous sessions.

You can't achieve your goals unless you know where you are going and you have a plan to get there. For this project, time was spent developing the strategy before diving into the deep end of the detail. Let's Go has proven to be very effective in communicating the JTZ message to our stakeholders. Additionally, as the coordinator of JTZ, I have continually referenced our strategy with stakeholders to ensure that we are still aligned. With a clear vision of our future state, we could confidently present to our leadership what it was that we needed from them in order to get us there.

Developing the strategy using the Let's Go model was not complex or time consuming. As the sign post on the Let's Go picture indicates, we started first with understanding where we were (the current state) and where we wanted to go. This was the starting point in building our high-level strategy. One of the most important questions that we asked was 'why' are we embarking on this change journey? Once we understood the 'why', the rest of the plan fell into place.

It was very important at the early stages of the project not to be lured into managing all the small details. It is the 'what' category of the Let's Go model where you may be tempted to do just that. If you go down this road, you will lose sight of the big picture and will never get over the finish line. Be patient. Finish off your strategy before jumping into the tactics (how).

If you don't communicate your change effectively, it won't get off the ground. This is why it is so important to know who your stakeholders are and what is important to them. You must develop your communication strategy with a clear understanding of what makes your stakeholders tick. Principal 3, Customers are the reason suppliers exist has, guided us when strategising over our customers.

To steer our company, our leadership needed to assess whether or not it was the right time to introduce a new idea and change initiative. There are competing priorities that must be managed. Safety is our number one priority, which makes the decision to support the initiative somewhat easier. However, a well thought out strategy will ensure that our valuable and finite resources are being used effectively and that we will achieve the results that we desired as a future state.

There are a few additional components that contributed to the successful implementation of the JTZ at In Situ. Our leadership wanted to be directly involved in this initiative. Instead of hiring a consulting firm, our leads were trained to be facilitators and were involved in the delivery of the five-hour sessions. There are now over 45 trained Facilitator Leaders in place and over 760 employees and contractors have been through the program. Through surveys completed at the end of each session, participants consistently commented on how important it was to see our senior leaders engage with them regarding safety.

(continued)

Suncor case study *(cont'd)*

The make-up of our pod structure directly supported our JTZ strategy. Each pod had one facilitation slot that was saved for an 'up and coming' employee. These sessions were an ideal opportunity for employees of all levels to volunteer to be part of the facilitation team. It is never a tough sell to get people to volunteer to be part of a JTZ pod. In fact, I have people from all over In Situ asking to be considered for future pods.

The leading indicators around employee engagement and empowerment, as well as safety performance indicators, are positive and trending in the right direction. It is currently under consideration that the JTZ program at In Situ may go company-wide. We are well on our way to developing and sustaining a high performance safety culture.

Figure 5.2: strategy on a page for the Let's Go change plan at Suncor

Journey to Zero	Strategy

Journey to Zero

Journey to Zero (JTZ) — In Situ Design

What's happening? Mike MacSween and his senior leadership team have sponsored the initiative to facilitate Journey to Zero sessions at In Situ. These sessions will be delivered by leadership at Firebag, MacKay River and Calgary.

The sessions have been initially developed by a task force that included GM Maintenance, Manager — Business Services and the EH&S Manager. They worked with a DuPont representative on the session material.

Strategy

Objective

Through the facilitation of JTZ sessions, employees will understand the overall concept of personal safety, environmental safety and process safety.

Where are we now?

Current state — At In Situ, we have a large number of new employees joining us. Some may have had attended some JTZ sessions, others have had no JTZ training. There still is a mentality that accidents can happen and we could not have done anything that would have prevented them from happening. We are still experiencing too many repeat incidents. Investigations are not at the level where we would see improvement.

Where are we going?

Future state — each employee (and large contractor/TAMS) member:

- develops an appreciation for Journey to Zero
- is immersed into the our Safety Culture through the belief that all incidents are preventable
- knows of the safety systems available and how to use them to move towards excellence in safety

The sessions will be facilitated through a pod structure. A member of the core leadership team will act as the lead, along with three members of the extended leadership team and an elite trainer facilitator (ETF). This structure allows for maximum flexibility in shift schedules, while ensuring that the pod members work together and are responsible for their specific sessions. Roles and responsibilities have been established for the pod members.

Sustainability of the JTZ sessions is infused into the strategy. Each pod starts off by facilitating to a group of supervisors and managers. Time during this 'test' session is spent obtaining valuable feedback meant to enhance future sessions. Additionally, once the pod has completed their designated sessions, a review meeting is held to discuss what worked well and what needs to be changed. During this review meeting, the lead of the next pod is expected to attend, thereby capitalising on first hand 'lessons learned'. Feedback from JTZ participants will be captured through a short survey handed out at the end of the session.

The JTZ strategy calls for a continuous improvement approach. The challenge is keeping JTZ front and centre and having it not become stale. Through electronic boards located all around In Situ, comments from the JTZ participants are captured. On a quarterly basis, a stewardship report will be compiled and reviewed by the Senior Leadership team.

Sandra Kendel,
Senior Process Improvement Specialist

Why are we going?

We are going on the journey to zero incidents to ensure that everyone goes home safely every day. By zeroing in on safety, employees will operate our facility with excellence.

What do we have to do to get there?

1 JTZ is all about leadership, at all levels of our organisation. The JTZ effort is not seen as a 'flash in the pan' effort, but one that is continuously communicated.

2 The In Situ Leadership team will be facilitating all of the JTZ sessions.

3 Pods are made up of a lead and three members of the Leadership Team. These pods are supported by an elite trainer facilitator (ETF).

4 Each pod will facilitate four sessions, with the participant numbers at 15–20 per class. The session is four to five hours in length.

5 JTZ communication is assisted through the electronic board strategically located throughout In Situ.

6 Registration for the sessions is coordinated through the admin assistants.

Who is involved on the journey?

- Trained pods consisting of the Leadership Team and the ETF Community.

- Initial focus is to train the new employees and work through the remaining workforce.

- JTZ sessions are for current employees and large contractors' employees.

- Training and updates to the JTZ material is coordinated through Operational Excellence.

When are we going?

The first 'test' session was 12 March 2009 and will continue until the entire In Situ workforce has gone through the JTZ session.

Linking current and future attitudes and behaviours to results

When most people think about change, they naturally think about changes in terms of results. This is especially true in the case of business change, where improvement objectives are commonly measured in terms of decreased cost, increased efficiency, increased market share, increased sales volumes, or increased revenue and profit. The improvement or performance goals of an organisation are often expressed as results.

How are changes in results planned and achieved? The answer: only through changes in the behaviour of an organisation and the people within it. A newly designed business process is only a concept until the people who execute the steps in the process are doing things differently — faster, better or in a more coordinated way.

The ability of an organisation and its people to implement sustainable new behaviour is, in turn, dependent on the attitudes, mindsets, skills of the people, and the supporting structure and systems of the organisation. These characteristics form the basis of an organisation's culture, and the individual beliefs and values of its people.

The process of the Let's Go change model

Complex change initiatives can be managed using the five process steps of the Let's Go change model shown in figure 5.3. The process will be unique to each change initiative.

Figure 5.3: the five steps in the process of the Let's Go change model

	Steps				
LET'S GO!!	Plan	Prepare	Implement	Review	Improve
Where (From and to)					
Why					
What					
Who					
When (Start and finish)					
LET'S GO!!	Provide **Leadership**	Engage, enable **Empower** people		Gain **Commitment**	Develop **Ownership**
	Motivators				

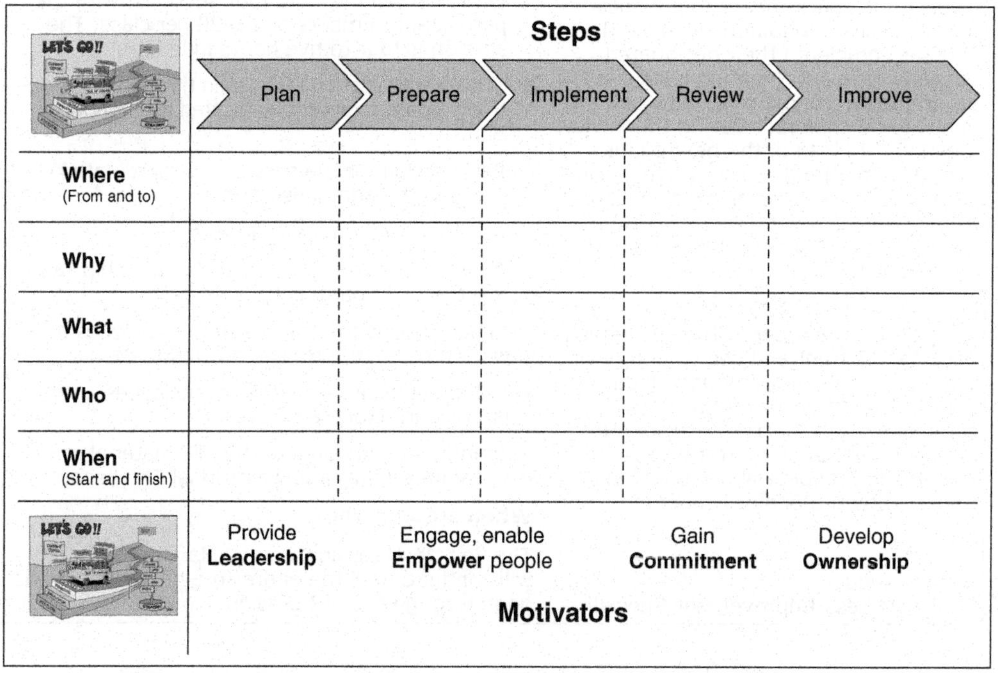

The five process steps involved in managing a change process are:

- *Plan* — Where are we now; where are we going to; and why is the journey important? Document the strategy and compelling value propositions for change. Is there a sense of urgency, a burning platform, catalysts for change, a clear vision and visible leadership to manage the change? What are the activities involved, who will lead and when will the process start and finish?

- *Prepare* — How ready is the organisation for change? Who is impacted? How will you engage, enable and empower people to gain their commitment and develop ownership for the change process? What are barriers to implementation? What is the transition plan? What is the impact on organisational structure, systems and processes. Who are the change champions who will support the leaders and the change initiative?

- *Implement* — Lead, manage and monitor the steps involved in transition and startup. Secure quick wins and consolidate on early success. Engage open, accurate and timely communication and information sharing. Empower and support the change champions. Celebrate success, both stretch and breakthrough improvement.

- *Review* — Measure progress and performance. Have leading and lagging, financial and non-financial measures in place. Celebrate success and implement the lessons learnt into the next phase of change consolidation. What has changed? What is different? How have people accepted the change?

- *Improve* — Implement continuous improvement processes. Continue to engage, enable and empower people on the improvement journey. As appropriate, re-engage and/or revitalise the plan, prepare, implement and review steps. How well has the change been anchored?

At each of the process steps, the Five Ws need to be asked: where (from and to), why, what, who, when (start and finish). Four prime motivators or drivers also need to be considered when managing change, and these are also relevant for relationship managers. They are:

- *Leadership* — Effective leadership is critical to setting, supporting and fulfilling the vision and mission for change. Relationship managers will play a prominent direct and indirect role in this process. Leadership styles will be situational and vary according to need and circumstance.

- *Empowerment* — Engaging and enabling people with knowledge, skills, attitudes and clear responsibilities to embrace change and play an active role in the change process will be the basis for empowerment. Empowerment improves initiative and motivation, leading to better quality decision making and more effective problem solving.

- *Commitment* — Commitment involves a pledge, promises or obligations by individuals to a specific course of action for which they are directly accountable. Commitment is manifested as unwavering support and buy-in.

- *Ownership* — Developing ownership is about people and teams successfully handling responsibilities, tasks and accountabilities as if they were of their own design or creation. Ownership engages hearts and minds and will generate passion as well as purpose. Without the three previous motivators there is little chance of developing ownership.

Linking corporate culture and leadership

There is a story that an IBM mid-level executive called a meeting with their boss to explain that they had just made a $10 million dollar mistake. 'I suppose you're going to fire me?', was the summary statement from the mid-level executive. The boss answered, 'No way. We have just invested $10 million in your education. Why would we want to fire you?'

Corporate culture can be defined as the deep-rooted values, beliefs and underlying assumptions of an organisation that determine how it functions and how it interacts with both its internal and its external environments. It is the heart and soul of the organisation, its personality and character, and is often spoken about in the context of 'the way we do things around here'. It is what causes people to think the way they do, say the things they say and do the things they do when involved with the organisation, and its customers and suppliers. Every company and organisation has a culture, be it strong or weak. It will be based on many factors, ranging from the history of the organisation, values, beliefs, symbols, myths and heroes to the influence of the national and regional culture. At its heart, corporate culture is about people and is palpable at all levels of the organisation.

There is a direct link between corporate culture and leadership. Strong, successful organisations have distinct corporate cultures and produce strong leaders, not only at the top but also throughout the organisation. In their book *Corporate culture and performance*, Kotter and Heskett analyse the nature of this relationship.

When describing how the cultures of the better performers had influenced their results, the interviewees often referred to qualities such as leadership, entrepreneurship, prudent risk taking, candid discussions, innovation and flexibility. Those were seen as cultural traits that helped firms do well in a changing business environment. In other words, they saw a causal link going from cultures that value leadership and the other qualities mentioned above to superior performance.

Kotter and Heskett[5]

Culture survey

To better understand corporate culture, try the quick self-assessment culture quiz in appendix B. It is built around 10 essential aspects of corporate culture.

Typically, cultural change is a lag indicator in that it takes time to initiate, implement and see results. Cultural change for an organisation can be a desired future state in itself. Culture is the most difficult and challenging, but the most important and rewarding area to tackle in relationship improvement because it goes to the heart of people and their values. Use the strengths and weaknesses and the areas of alignment and misalignment in the culture quiz results to identify opportunities for improvement.

Three critical relationship roles

Whether internal or external to the organisation, if customers are the reason suppliers exist (Principle 3), everyone in the organisation has relationship management somewhere in their job description. This further reinforces the argument for developing relationship management as a core competency. There are, however, three specific roles that will become increasingly important in the delivery of business strategy and the improvement of business relationships. They are:

- chief relationship officer (CRO)
- 0 to 10RM relationship facilitator
- 0 to 10RM relationship manager.

These three positions are central to a community of practice business model that entails empowering high impact individuals in different roles, top to bottom, in the organisation. Their aim is to share knowledge, implement best practices, create and facilitate breakthroughs, and attack common problems with creative solutions. This applies internally between functions and across external interfaces with customers and suppliers. They work across the formal and informal organisational structures and networks with clear goals, explicit accountability and executive oversight. Supporting leaders and providing specialist expertise, the community is an accelerant for generating employee engagement, commitment and ownership. By empowering individuals who are aligned with the vision and values of the organisation, their collective efforts become catalysts for constructive change. The community of practice is a vehicle for business improvement and organisational transformation.

You no longer evaluate an executive in terms of how many people report to him or her. That standard doesn't mean as much as the complexity of the job, the information it uses and generates, and the different kinds of relationships needed to do the work.

Peter Drucker,[6] author, management visionary and pioneer

Chief relationship officer

Along with the chief executive officer, the chief operating officer, and the chief financial officer, the chief relationship officer (CRO) role is critical in managing the health and wellbeing of the organisation's most important assets. It is a direct recognition from the organisation of the importance of people and relationships. The key aspects of the CRO role are:

- To lead and steward the health and wellbeing, approach and performance of the organisation's most important internal and external relationships, including but not limited to customers, suppliers, stakeholders, regulators, employees, internal functions, business units and departments.

- To envision and guide the organisation's relationships strategy and ensure its alignment with vision, values and broader business goals.

- To challenge the prevailing paradigms and implement best practices in relationship and change management.

- To develop high performance relationship management as a core competency for the organisation and therefore a point of difference and demonstrable source of competitive advantage.

- To use high performance relationship management to deliver business goals, continuous and breakthrough improvement.

- To develop and sustain organisational relationship and change management capability within the organisation.

- To coach and mentor relationship facilitators and relationship managers to achieve their potential and maximise the value added for the organisation.

The CRO provides the top-down executive commitment, support and leadership that enables the bottom-up engagement from all the relationship participants and stakeholders to be sustained.

The 0 to 10RM relationship facilitator

The role of the relationship facilitator (RF) is to directly support and partner with leaders, relationship managers and teams in the delivery of the organisation's broader corporate goals and objectives. Sydney Water Corporation RFs call them partnering facilitators, as they play a critical role in the organisation-wide, internal, transformational partnering initiative. Suncor Energy in Canada call them change champions to support major change initiatives. China Light & Power (CLP) in Hong Kong and Gen-i in New Zealand have elite trainer facilitators and coaches in relationship management. ICI Australia in the mid 1990s, as part of paradigm shifting workplace reform, used the title workplace facilitators for specially trained union shop stewards.

The role combines high level facilitation skills, relationship management experience and knowledge, along with existing skill sets and job accountabilities that individuals brought to the role. In addition, RFs take on the roles of coach, mentor, provocateur and trainer as required. They have a freedom of speech, movement and association that few others in the organisation enjoy. Empowered by executive authority and oversight, there are few more exciting, challenging and rewarding roles in business today. High level RF role responsibilities are as follows:

- They support and partner with leaders and relationship managers in the delivery of organisational goals and business strategies and transformational organisational initiatives.

- They engage in preparation, delivery and follow-up activities associated with the effective facilitation of a wide range of high impact workshops and meetings using specific relationship management models and facilitation tools and techniques.

- As facilitators, they are neutral, trustworthy, active listeners focusing on the process (that is, tools, methods, group dynamics) enabling and helping others to focus on the content (that is, the task, subject matter, the decisions made).

- They train, coach and mentor others in the practical application of high performance relationship management models and tools.

- They deliver, directly and indirectly, sustainable value, internally or externally, for customers, suppliers and stakeholders.

- They deploy general facilitation skills to assist and accelerate business improvement, such as assistance in defining group goals, workshop facilitation, effective meeting management, decision making and problem solving, conflict resolution, innovation, continuous and breakthrough improvement.

- RFs are provocateurs who challenge the status quo and prevailing paradigms that impede positive progress and improvement, namely, those mindsets, behaviours, attitudes, practices, policies and procedures that often go unchallenged and that have a negative impact on change and relationships. That can include 'this is the way we do things around here'; 'but we have always done it that way'; and 'we tried that already, years ago, and it didn't work'.

- They are adaptive to change and creative when it comes to solving problems and creating opportunities.

- RFs are storytellers who have the ability to engage others to tell their own stories.

Ed's story

Good candidates do not always present as the obvious first choice. This is a story about Ed, an operations veteran of 28 years in the oil and gas industry, 20 years of which were spent as a tough, straight-talking union shop steward. For the last eight years he had been taken off the tools and been given a supervisor's role. Ed had developed a reputation as a hard-nosed, opinionated leader who had zero tolerance of unsafe work practices, be they at the management or shopfloor level or anywhere in between.

Through a combination of recommendation by a work colleague and a genuine interest in the role, Ed found himself on a relationship facilitator training program. I was fortunate enough to be the trainer in this instance. Ed was the last person around the table to introduce himself to the group and overview his personal objectives for the five-day program. He looked me straight in the eyes and said, 'I want to learn how not to be a saboteur'. That was it. A simply remarkable statement. Prompted by some group questioning he went on to add that he had been a combatant for most of his working life and he was sick of it. 'Not enjoyable, not constructive and not sustainable and there has to be a better way' were his words.

Ed was a stand-out participant in the program and subsequently went on to make a huge difference to the business. His ability to positively engage shopfloor staff and management alike, to tell stories and lead people to better places turned out to be a remarkable skill set that added great value to his own personal growth as well as the business.

I must confess that, on paper, I would not have initially nominated Ed as a program candidate. Thinking relationship facilitator was solely a role for the highly educated, self-confident, commercially experienced and ambitious up and coming senior manager was a false assumption. Ed's background, training and reputation did not fit my mould as a high performance relationship facilitator. Ed turned out to have genuine empathy, self-awareness and self-confidence, and intuitive judgement that made him an ideal facilitator. It turns out that emotional intelligence (EQ) is as important as intelligence quotient (IQ). It also happened that he was a very talented poet in his spare time. Who would have thought? It is amazing what can happen when good people are given an opportunity and the freedom to utilise often hidden natural abilities.

Conditions for relationship facilitator success

Three critical success factors are lessons learnt around effectively supporting relationship facilitators:

- Executive oversight, commitment and support.

- Support from the immediate leaders and managers to allow the time and opportunity to make a difference and evolve competencies.

- Implementation of effective leadership and governance structures that support the community of practice model and the delivery of business improvement.

The 0 to 10RM relationship manager

Relationship managers offer opinions, influence others towards win–win solutions and rarely take a neutral position. Relationship managers are often embedded in the content, working with the strategies, tasks, subject matter and decisions made. They are directly responsible and accountable for the health and wellbeing of the relationship approach and performance.

0 to 10RM relationship managers are central players in relationship rescue, relationship improvement and relationship transformation initiatives at strategic and operational levels, internally and externally to the organisation. 0 to 10RM principles, models and tools have been specifically developed to support both the relationship manager and relationship facilitator roles. Both roles require complementary skill sets and competencies that together have a compounding value-adding effect.

The difference between the relationship facilitator and relationship manager role lies in process versus content. Relationship facilitators lead the process steps rather than work in the content, and they pride themselves on their neutrality, whereas relationship managers are typically immersed in the content, are passionate about their opinions and take direct ownership of delivering relationship objectives. There is always the opportunity for interchangeability across both roles, whereby a relationship manager in one area of the business can be a relationship facilitator in support of another part of the business.

Appendix B provides a survey on general high performance relationship manager competencies. Take some time to complete the survey to better understand your strengths and opportunities for improvement surrounding relationship management skills.

Getting internal relationships right

One of the biggest impediments to the delivery of external strategy is the quality and performance of internal relationships. 'Over-promise and under-deliver' is the classic trademark of an organisation that does not have its internal relationships right. An example would be a delighted salesman returning with a bonus-winning order only to find that operations can't deliver. Some explanations from operations might include:

- the order is not in the forecasts

- it can't be fitted it into the schedule or plan

- you haven't allowed enough lead time or notification

- we should have been told earlier or at least been given a heads up

- the spec is too tight

- we don't have the parts or raw materials

- there's a cost-efficiency focus that conflicts with product or service innovation.

Internal alignment is crucial in delivering on external commitments. Organisations must put their own house in order before, or in parallel with, forming high performance relationships with external customers and suppliers.

The challenge awaits throughout the organisation, but particularly for management, who must have the courage and foresight to instil the spirit of discovery and change, not only in the people they lead but also in themselves. Good leaders and managers must actively participate in the relationship-building process and not abdicate this responsibility under the guise of delegation and empowerment. The obsolescent view of imposing control to achieve compliance in most cases will be counterproductive. The alternative is to elicit commitment, participation and ownership through effective leadership and trust building.

RAD analysis and relationship strategy maps (RSMs), Let's Go action plans and other 0 to 10RM models and tools are as appropriate internally as externally. Internal relationships need to have a willingness and capability that at least matches the relationship approach of the most strategic customer and supplier relationships — otherwise, the ability of the organisation to fulfil its external strategy will be fundamentally compromised.

Summary

- Relationships are all about people. High performance relationships need to be enduring and successful beyond the life of key people.

- Principle 4, The right people doing the right things, in the right way, at the right time for the right reasons will deliver the best possible outcomes, lies at the heart of the Let's Go change model.

- There are four different types of people on the Bus of Change: innovators, early adaptors, followers and saboteurs. Identify these people and understand how they can contribute to relationship improvement. The four types of people can be found anywhere within or external to the organisation.

- Innovators and saboteurs share three common qualities: both groups are stubborn, passionate and unreasonable. Saboteurs have the potential to be innovators.

- The Let's Go change model involves mobilising the Bus of Change on the journey of improvement to the desired future state. It is a call to action.

- The Five Ws within the Let's Go change model apply at the strategic and tactical or operational levels: where are we now and where are we going (from current state to desired future state); why are we going there; what do we need to do to get there and how do we need to do it; who is involved and impacted; and when do we start and finish, and what are the milestones on the journey?

- Understanding cultural alignment is critical in the development of high performance relationships. Typically cultural change is a lag indicator in that it takes time to initiate and implement. Diversity and cultural differences may be leveraged to advantage.

- Culture (the way we do things around here) goes to the heart of the organisation, its people and their values. It is what sustains good relationships.

- Passionate champions lead the journey toward sustainable success in organisational relationships. The three critical roles to develop are chief relationship officer (CRO), as sponsor of the change; relationship facilitator; and relationship manager.

- Developing communities of practice around relationships, relationship managers and relationship facilitators will positively support organisational transformation.

Part C

JOURNEY MANAGEMENT

CHAPTER 6
THE RELATIONSHIP DEVELOPMENT
CURVE

While it is true that the relationship approach is the means to an end and not the end in itself, the journey is where lessons are learnt, relationships are built, problems solved and insights gained. Yes, the end goals are important. Like bookends, the two anchor points, the current state and desired future state provide the boundaries of the process. They provide the purpose and focus around which the journey can be taken. But the real joy is in planning for and taking the individual steps, and sharing the pain and the gain along the way. This is what journey management is all about.

Listen to anyone who has climbed Mount Everest, sailed around the world, run a marathon or travelled across continents, and they will rightly delight in having achieved their end goal. However, it will be the journey highlights and lowlights around which the stories are told and new solutions to old problems are found. These moments of truth provide the inspiration, courage and experience that enable us to set sail for new horizons.

Just like relationships in life, business relationships go through stages and phases. They evolve over time and are a rollercoaster ride of emotions, effort and outcomes. There are few straight lines in life or business. Figure 6.1 (overleaf) outlines the 0 to 10RM relationship development curve, which shows the connection between time, the quality of relationship outcomes and the different phases engaged throughout the improvement journey. The timeframe and rate of change are indications only. They depend on the people involved, and the internal and external business environment.

The 0 to 10RM relationship development curve is not an esoteric or theoretical model, but a practical tool developed over many years of observation and experience. There is also a direct link between the Let's Go change model (discussed in chapter 5) and the journey management concept associated with the relationship development curve.

Figure 6.1: the relationship development curve

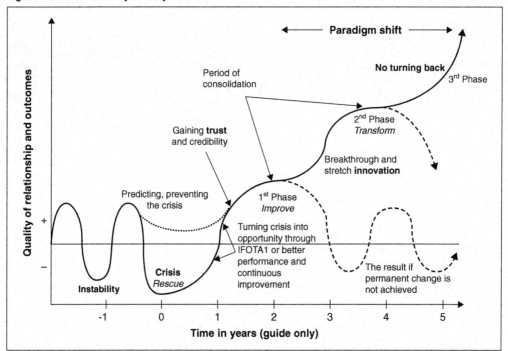

The stages in the development curve

There are five main stages on the relationship development curve:

- instability
- crisis
- trust
- innovation
- no turning back.

Instability

On the relationship development curve, time zero refers to the point at which there is a significant or fundamental shift in the relationship. Business relationships can go on for years in varying states of instability before time zero. They navigate through peaks and troughs, of human behaviour, quality, service and schedule performance, and improvement requirements or expectations, the highs and lows of commodity price cycles, and the uncertainty of exchange rate fluctuations. Theme 1, the

0 to 10RM Matrix (figure 1.4 on page 10), shows that if the relationship is not where it needs to be in this fluctuating cycle — that is, the current state and desired future state for the relationship differ — there will come a time when a step change or fundamental improvement must occur.

This impetus for change or improvement around threats or opportunities could be related to quality, cost, service, schedule, commercial and market-related issues, processes, systems or people. There may even be a related crisis or tipping point that threatens the very future of the relationship. Denial and resistance to change may manifest as confrontational, adversarial, protective and defensive attitudes and behaviours to getting things done. In some cases, the relationship parties drift from one problem or complaint to another. They are often blissfully unaware of the alternatives. Between these problems or complaints are periods of relative calm, but little progress. Intermittent periods linked to the arrival of new people, new products or services or technologies, reorganisation or process re-engineering give short-term respite from the brewing crisis. This instability might continue for years.

Crisis

Natural disasters are good examples of where a crisis can affect relationships in both our business and personal lives. Earthquakes in Chile, tsunamis in Indonesia and Japan, fires in Greece, erupting volcanoes in Iceland or floods in Australia and Brazil — the one common positive theme that arises out of crisis is the marshalling of community spirit. Why do we so often need a crisis, such as a natural disaster, to show our true character as human beings and to generate a sense of urgency? Why is it so easy to adapt to and adopt change in the face of an overwhelming, often life-threatening, crisis? The short answer is that there is a focused set of common goals around which it is to everyone's benefit to cooperate and collaborate. Safety is the obvious first and over-riding concern; assets and property are second priority. Staying where you are, maintaining the status quo, is not an option in these situations. As individuals, teams, social networks and communities we become those right people doing the right things at the right time for the right reasons.

What we see arising out of these crises are the human stories of bravery, tolerance, giving, caring and community spirit, of people working together, finding new ways of doing things and making new plans in the face of seemingly insurmountable difficulty. Ordinary people do extraordinary things in a crisis. Life-saving relationships become life-long relationships.

Yet often, when the immediate crisis has passed, attitudes and behaviours revert back to pre-crisis settings. Even to the extent that finger pointing and blame set in. Adversarial politics, and protect and defend behaviours surface; self-interest based deals are done; and all too often the bipartisan relationships formed during the crisis break down.

Returning to the relationship development curve, let's assume that time zero represents the lowest point in the relationship development. This is often the point of crisis, from which there is no return. From this time on, the nature of the intervention will determine the path of the relationship into the foreseeable future. It may be that the business has already been lost and there is one last opportunity for the relationship to survive. Not all relationships start in this manner; some could be situated at a point to the right of time zero on the development curve.

Smart organisations will pre-empt the impending crisis through preventive, predictive and design-out initiatives. The broken line running above the crisis point represents a short circuit on the development curve that provides a bridge over the crisis point at time zero.

The challenge for the relationship parties is how to generate a sense of urgency in advance of the imminent crisis, before the relationship is yet again propelled into a trough of hostility and concomitant poor performance. The perception at the high point, before the fall into crisis, is that the relationship has never been so good. Telling people at the height of their success that they must change, in some cases fundamentally, or risk failure is not an easy sell. Some in the organisation, including senior managers, have a strong investment in the prevailing paradigms. Their argument is often 'if it ain't broke, don't fix it'.

The fallback strategy to pre-emption and prevention is a well-managed crisis and associated rescue. Another perspective is that the real crisis may not be the depth of the fall into the crisis, but rather the slope and height of the climb out of it. In that case, a graceful exit strategy to a zero relationship type may be the more appropriate course than a stillborn improvement process that quickly reaches a point of diminishing returns.

A crisis or the threat of a crisis can be the ideal catalyst for positive change. If you want change and don't have a compelling reason or crisis, find or manufacture one! This, however, is a double-edged sword. Unless change is enacted out of the crisis, the environment, attitudes, practices and behaviours that caused the crisis in the first place are likely to remain and so the crisis is likely to reoccur. For example, the reforms to the banking sector born out of the 2008 global financial crisis (GFC) will be a genuine test of whether attitudes and behaviours have changed to a point where a similar crisis will not recur.

In terms of the relationship development curve, once the crisis point has been dealt with there is typically a phased roller-coaster ride of activity and improvement along the journey pathway. There is a clear connection here between Principle 5, Insanity is doing the same things and expecting different results (see figure 6.2), and the relationship development curve.

Figure 6.2: Principle 5—insanity is doing the same things and expecting different results

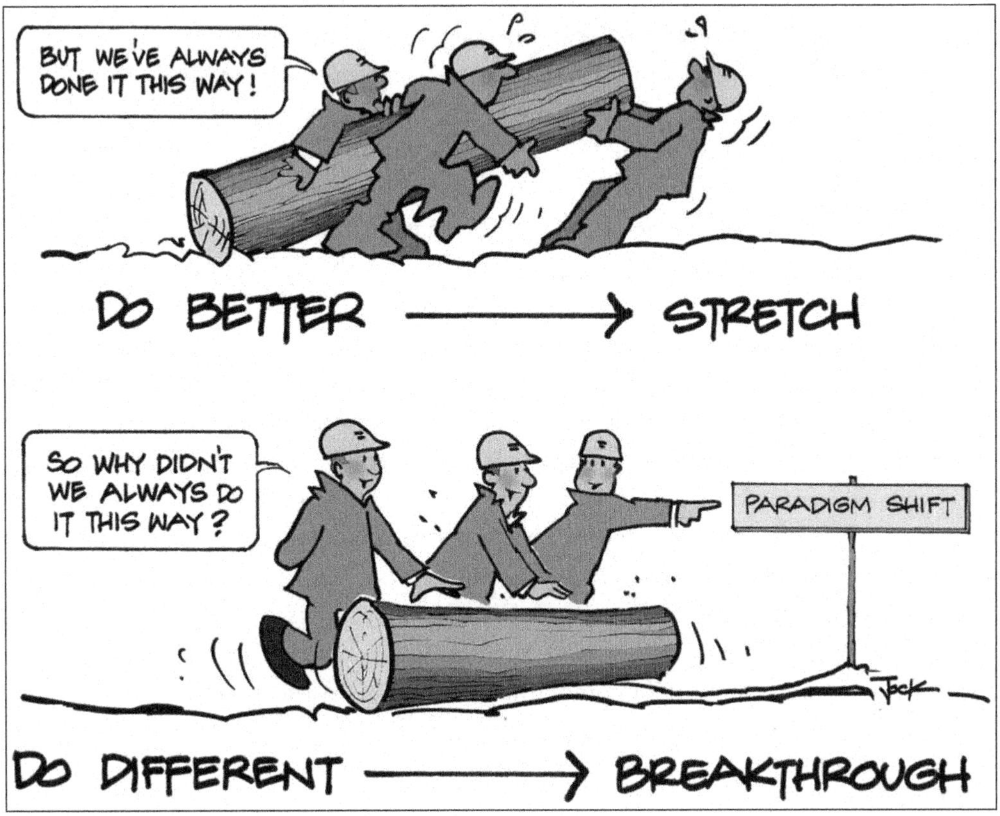

Trust

Building trust and credibility on the basis of meeting or exceeding agreed requirements IFOTA1 specification, and delivering genuine improvement to indicate that a real change has occurred, will take time. On the time scale shown in figure 6.1 on page 186, with the crisis point at year 0, the customer will notice a significant and measurable change at around 18 months. This is the first phase of improvement beyond time zero. The rate of change will continue for another six months before a period of consolidation is reached at year two.

From my experience, a point is reached where a pause in progress occurs. This pause allows time for reflection and consolidation, and planning for the next steps. This is often when the new paradigms are discovered and revitalisation, reset, recalibration and transformational initiatives are engaged. Just like elite athletes, people and business relationships cannot sustain continuous levels of peak performance.

Once on the journey, don't forget the followers and the saboteurs. The innovators may well want to push on over the horizon; while the saboteurs are setting up roadblocks and other obstacles to improvement. At this point the early adaptors can provide a critical link between the innovating pioneers and the followers. Turning crisis into opportunity can be a time of intense change in both attitudes and processes.

Some people need more time than others to get on board the Bus of Change. In figure 6.1, the relationship development curve, I have allowed six months for the period of consolidation. I should emphasise that this is not a period of stagnation. It will involve actions such as engagement with stakeholders, reflection, reviewing lessons learnt, framing, bedding down the change process and discovering new paradigms. Unfortunately, it is also the point at which apathy, complacency and arrogance can creep in, particularly if the process is not managed effectively. People sense they have done a great job and think that no more action is required. Strong leadership and clear vision are vital to avoid slipping back to the comfort zone or, even worse, back to before time zero with all its instability and variation. In figure 6.1, this is indicated by the broken line that begins just after year two. Creating a more viable, attractive and common vision for the future is paramount if the relationship parties are to jointly develop new rules of engagement. People, relationships and organisations, and their practices and beliefs, will not change unless the future represents a compelling alternative to the past. This will involve both continuous and breakthrough improvement.

Innovation

In figure 6.1, at two and a half years into the relationship improvement process and all being well, we start to enter the paradigm shift. All the easier problems have been solved and the associated opportunities implemented. New opportunities are being discovered and new problems solved through the new paradigm(s). The second phase of change in growth and development, between two and a half and three and a half years, moves the relationship focus of meeting current requirements to a goal of exceeding requirements and meeting future requirements. This is a transformational time, when innovation outweighs the importance of fulfilling requirements IFOTA1 specification. Meeting requirements IFOTA1 specification is now taken for granted. At this point of reinvention, relationships between customer and supplier are strong, broad and based firmly on a high level of trust and mutual commitment. This allows for best in class, best value innovation and next practice delivery of other high-level benefits.

Creation is given birth by those people in a perpetual state of enlightened dissatisfaction.

Robert Porter Lynch[1]

A period of consolidation typically occurs between three and a half and four years. The reasons for this might include changes in key personnel, time taken for followers to catch up, and the development and implementation of new technologies. Again, vigilance is required to identify pockets of apathy, arrogance and complacency.

No turning back

A new phase of change is entered into at four years beyond time zero. This will carry through to year five and beyond. The process is now virtually self-motivating and self-perpetuating, with the relationship robust enough to handle almost any circumstance. A genuine and sustainable high performance business relationship is now in place. The cycles of growth and consolidation will continue for as long as the relationship participants desire. It is my view that, once this phase has been reached, irrespective of the time involved, both customer and supplier have changed and improved so significantly that they are beyond the point of no return. Irrespective of what happens in the future, once the benefits and the trust have been experienced, there can be no going back to the way we used to do things or at worst the bad old days.

Success is not a place at which one arrives but rather the spirit with which one undertakes and continues the journey.

Alex Noble[2]

Applying the relationship development curve

If you think the 0 to 10RM relationship development curve is applicable to the relationship(s) you are involved in, or are developing, you many enjoy this exercise. You will find it an enlightening reality check. Ask yourself the following:

- Where does the relationship(s) currently sit on the relationship development curve? Substitute your own time scale if appropriate.

- What events (such as activities, initiatives, milestones) have occurred to get there?

- Where is the relationship going on the development curve, in what time frame, and what goals, strategies, activities and milestones will enable the relationship to get there?

- What is the next paradigm shift that will take the relationship to the next phase?

- If the relationship currently sits at or before the crisis point, what is the burning platform for change? How can the crisis be avoided?

Paradigm shifting

The futurist Joel Barker in his Discovering the Future Series, talks of paradigms as 'patterns of behaviour and the rules and regulations we use to construct those patterns'.[3] They are often spoken about as 'the way we do things around here'.

Barker further states: 'In almost all cases we measure our life's success by our ability to solve problems within our paradigms. In science, in business, in politics, in education, in our lives — changing a paradigm means fundamentally altering the way things are done.' Paradigm shifts can be big or small, but they are all fundamentally important for those people practising the prevailing paradigm. This is not only about the world being round and not flat, Earth revolving around the sun and not being the centre of the solar system, or being able to travel faster than the speed of sound.

You see things and you say why? I dream of things that never were and say, why not?

George Bernard Shaw[4]

Sport, for example, is laden with paradigm shifts. At the 1968 Summer Olympics in Mexico City, American athlete Dick Fosbury single-handedly revolutionised the sport of high jumping. He won the gold medal and set a new Olympic record at 2.24 metres by jumping over the bar backwards. His technique became known as the Fosbury flop — it was the opposite of the then prevailing western roll and straddle techniques, which involved jumping forwards or sideways. The Fosbury flop has since become the standard technique for high jumpers all over the world and it continues to enable the setting of new world records.

In business, a paradigm shift can be as simple as moving from an overtime work environment linked to re-work, system breakdowns or plant outages, to an annualised salary work environment based on improved productivity and reliability, like the one introduced at ICI in Australia in the 1990s (see chapter 3). Entering a two-way, open-book, performance-based, transparent and trusting relationship rather than trying to sustain a closed-book, low-trust, non-transparent relationship is still a paradigm shift for many organisations.

New paradigms and paradigm shifts are often the domain of Type 8 Partnering and Alliancing, Type 9 Pioneering and Type 10 Community relationships. Pioneers, such as Mohamud Yunus of the Grameen Bank, are changing the banking relationship paradigm with micro-finance; the Bill and Melinda Gates foundation is engaging a new paradigm around philanthropic capitalism and global philanthropy; Project Alcatraz is a new business model based on social entrepreneurship; Wikipedia, Linux and Firefox are open source communities; and social media, like all of these, is fundamentally changing the business networking landscape (see chapter 3). It is an irony that new paradigms are mindset-shifting and value-adding yet, when they are implemented, the response typically raises the question: why haven't we always done it this way?

Tyre paradigms

Paradigms shifts at the workplace level present themselves every day. For example, the new commercial paradigm of getting paid on the performance and effect of products and services, and not only their unit price or cost, is linked directly to new relationship behaviours and practices, and win–win outcomes. At first glance there is no better example of a commodity product than that of tyres. But in the words of Theodore Levitt 'every product and service is differentiable'.[5] The following examples look at tyres from a whole new perspective.

Flying

Companies supplying tyres for aircraft and getting paid on dollar per landing and not dollar per tyre needed to build better networks and relationships with everyone from pilots to maintenance and customer service teams to ensure maximum understanding and engagement around optimising tyre performance. The win–win in this relationship is softer landings, which make for happier customers, and an increase in the life and therefore the number of landings of the tyres. And that means lower total costs for the airline, and bigger margins for the tyre companies. Happy customers are then prepared to come back more often to travel, and may even be prepared to pay a price premium or better flying.

Trucking

Tyres for large tip trucks in the mining and resource sector are being sold on the basis of dollar per kilometre tonne and not on the basis of dollar per tyre. This change has required the tyre companies to build better relationships with road maintenance teams to improve off-road surface conditions to reduce maintenance costs and to cut tip truck downtime. In making this paradigm shift, the companies discovered that women often make better drivers of these large vehicles, which in turn reduces the wear and tear on tyres, giving them a longer life. Again, the result is lower costs, higher margins and happier customers.

Run flat

Rubber giants Dunlop, Michelin, Goodyear, Bridgestone and Pirelli have been working for many years with their suppliers and the automakers to develop run-flat tyres. These tyres don't deflate and can run for more than 100 kilometres if punctured, without the wheel trim being ruined. They are the next step beyond self-sealing tyres. These run-flat tyres are widely used in high performance and prestige vehicles, and BMW and Mercedes now regularly fit them as standard equipment. Other automakers that supply run-flat tyres on the vehicles include Audi, Daimler-Chrysler and Ferrari. The tyres sell for more than twice the price of conventional tyres, but it is not just the quality, convenience and safety as opposed to unit price per tyre that will make them successful. Goodyear has developed a run-flat tyre so reliable that the vehicle doesn't need to carry a spare — a key goal for all tyre manufacturers. This is a blessing for designers of sports cars and four-wheel drives in particular, because it frees up new design options and available space, and avoids the dilemma of where to put the fifth wheel — on the tailgate, inside the vehicle or underneath the body.

My point is that the impact of tyre development extends beyond the wheel trim and into car design, lower total costs and ultimately the added value for which the car owner is prepared to pay. Focusing on value, not unit price, provides benefits up and down the supply chain. For these new paradigms to be engaged, higher performing relationships need to be developed. The ongoing pressure to downsize, restructure, price cut and cost cut, however, often leads to an unbalanced focus on lowering supplier or service provider unit prices at the expense of innovation, total quality, total value and total cost improvement.

The value question

The value question is, What value is this relationship delivering for this organisation over the alternatives?

Any idiot can reduce a price by 10 per cent to become more competitive, but if you can offer an electric power transmission cable under the Baltic one year earlier than your competition, that is of tremendous value to the customer, and your competitor can't touch you.

Percy Barnevik, Chairman, Asea Brown Boveri (ABB)[6]

How do you keep the financial bean counters, nay-sayers, cynics and occasional saboteurs from cutting short your brave new world expeditions with their corporate stories on the relationship's credibility and overall net value?

The value for money question is one of the biggest challenges in sustaining high performance relationships. The fact that many senior executives are under great pressure to deliver continuous and short-term focused shareholder wealth and financial returns continues to adversely impact the quality of their long-term decision making. Is it any wonder that long-term strategic focus often succumbs to the pressures of short-term opportunism? The following true story has achieved legendary status.

The value question and the new CEO story

The new CEO of the customer partner organisation in what had been widely publicised as a very large, highly successful Type 8 Partnering and Alliancing relationship called a meeting at short notice with the two customer and supplier alliance managers. It was two years since the signing of the agreement, so the honeymoon or transition period was well and truly over. The alliance managers were somewhat apprehensive and wary as their mail had been telling them that the new CEO was not a big fan of long-term, collaborative relationships. His reputation from his previous roles was one of a typical tough guy, hard-nosed, hard-dollar and bottom-line driven. Relationships for him were for families, friends and pets, and the company alliance didn't come into any of these categories.

They arrived at his office and true to form he got straight down to business. 'I've heard a lot about this alliance relationship but little of the detail. So what value is it delivering for this organisation over the alternatives?' he asked, clearly in no mood for casual conversation.

Caught more than a little off guard, like deer stunned in headlights, their response, was 'Really well'. They might as well have added, 'thank you for asking and caring'.

(continued)

The value question and the new CEO story *(cont'd)*

The CEO countered, 'No, I'm interested in the detail. How is this relationship delivering value over the alternatives? Because I think I have some better ideas!'

It's difficult to think quickly and clearly when your heart has just skipped a beat and a state of panic has set in. To cut a long story short, while the alliance managers had plenty of information to give, good stories to tell and ultimately good results to share, the data was not in a form that was readily presentable. To his credit, the CEO gave them two months (that is, two alliance management team meetings) to successfully demonstrate the answer to the value question.

In response he was provided with a simple KPI scorecard that linked achievements directly to the alliance charter objectives and value propositions for the relationship. The alliance managers were able to demonstrate sustainable value past, present and future through historical and projected baseline analysis, and best practice reliability and productivity as measured against known industry standards.

As the alliance managers later told me, 'We learnt a lot that day and in the subsequent months. The previous CEO had initiated the alliance and we had total support, as much on faith as on substance. Over the first two years we had focused more on performance than on measurement and had been effectively left alone by external forces. We knew the results and improvements were there and that the alliance was delivering superior performance over the alternatives, but the data was not fully visible or in a form sceptics or outsiders would readily accept. We naively thought that the support would automatically continue unchallenged. We were wrong. What followed that initial meeting with the new CEO was an 18-month program of systems and process development to not only ensure good data and measurement but also drive improvement. As it turned out the value question was the right question to ask at the right time and the new CEO is now as big a supporter of the alliance as was his predecessor.'

You may be wondering what alternatives the CEO had in mind. They were nothing more than a couple of back door 10 to 15 per cent discount offers from competitors of the supplier partner for similar services. While attractive on the surface, they were nothing more than low-ball bids with a short-term perspective. As the ideas were never accepted, their effect on quality and service can only be imagined!

It is fair and reasonable to ask the value question, as the CEO did, and it deserves a clear and succinct answer. The value question provokes three sub-questions:

- What is the value being delivered?

- What are the alternatives?

- How do they compare?

High performance relationships, be they Trading, Major, Partnering and Alliancing or any of the other relationship types, must prepare for and be able to answer the value question and be able to demonstrate that the relationship is delivering sustainable value over and above the alternatives.

Strong high performing relationships and the associated parties do themselves a great disservice by not effectively measuring the ongoing sustainable value being delivered. Whether this involves external benchmarking against competitors, other organisations or industry standards, or absolute improvement against agreed baselines, measurement is one of the more important aspects of high performance relationship management. Some essential insights into the value question include the following:

- Ensure that the relationship not only survives but also thrives and grows beyond the life of key people. This involves ongoing education, induction and communication to ensure newcomers, irrespective of their role, are made familiar with the relationship approach being taken and the value propositions that underpin that approach. Do not under-estimate the power of common understanding, common language and common practice around relationship management generally.

- The importance of effective stakeholder management and their associated buy-in and commitment should not be undervalued. Keeping key stakeholders involved and informed may remove or prevent ongoing misalignment and friction.

The 10 solutions to the value question typically comprise one or more of the following activities:

- Benchmark against accurate historical costs or baselines and associated forecasts.

- Benchmark against parallel activities or relationships (internal and/or external).

- Collect peer review or anecdotal evidence.

- Conduct independent or expert third-party reviews or audits.

- Compare actual versus forecast relationship improvements.

- Benchmark against national and international standards or available competitive or industry data.

- Use best practice benchmarking.

- Seek industry recognition and awards.

- Establish the triple bottom line (business, social, environmental) benefits that have been delivered.

- Test the market or use go to market strategies (jointly or separately).

Crisis and opportunity management

A crisis typically brings out either the best or worst in people. In situations of war, life adventures and sport, as well as business, how many times have we seen victory being snatched from the jaws of defeat? How often have we seen teamwork, loyalty, skill, courage and determination lead to success and be rewarded by enduring respect and trust from friends, family and peers? Relationships in business are no different. The quickest and probably the easiest way of developing a good relationship with a customer or supplier, based on all the right qualities, is to solve an immediate problem. It is often the centrepiece of a good relationship rescue. The bigger and more urgent the problem, the better, especially if it is solved in a quality way, namely:

- Define the problem.

- Put a quick fix or temporary solution in place.

- Determine the root cause of the problem.

- Implement more permanent or longer term corrective action(s).

- Evaluate and follow up for continuous improvement.

Permanent or long-term solutions to seemingly insurmountable problems are not only impressive, they also generate confidence in capability and ongoing loyalty, and provide the ideal platform for building trust. Unfortunately, solutions to problems tend to be treated as one-off events rather than strategic opportunities, and so there is a feeling of relief rather than excitement when the problem is resolved. Organisations need to build a culture where employees actively seek out crises and problems, whether precipitated by their own organisation or another. Problems and crises, and, therefore, opportunities, occur every day in areas within and outside the immediate domain of the products and services involved in a business relationship.

There are areas of expertise and talent throughout every organisation that may never be exploited or even thought to exist. Create an inquisitive environment and train your people to seek out and solve problems, turning crises into opportunities. Establish training programs in communication and problem solving, observation and questioning, analytical, technical and IT skills, teamwork and leadership. This is not about crisis management, but rather crisis recognition. Minor and major problems occur every day, but most people are not trained to identify and act on the indicators. Take advantage of these unique opportunities to show your stuff, consolidate the relationship, and build loyalty and trust, and in turn, competitive advantage.

Condition monitoring story

I like the story of the relationship manager who receives a phone call from a customer partner having operational problems with loud, irregular noises coming from a large gearbox and bearings on their production line. They are key components of critical in-line machinery, so their failure would effectively bring the whole production line to a stop. There are no spares on hand and there is also a six- to nine-month lead time for replacement parts to be shipped from Europe. The customer had few diagnostic skills on site to determine the nature and extent of the problem. With the production line running at half rates, the threat of cost blowouts, declining stock levels, production downtime and lost sales was immediate.

Unrelated to the products and services being sold, the relationship manager discovered that a genuine centre of engineering excellence existed within his own organisation in the area of condition monitoring and vibration analysis of gearboxes and bearings. The team's forte was understanding the nature of gearbox and bearing problems when they arose and predicting with considerable accuracy what the time to failure was.

Within 24 hours the engineering team had visited the customer's site. Initial tests were done and a plan agreed to, namely, immediate corrective action and longer term preventative and predictive maintenance. It turned out that there were major problems requiring attention. However, the time frame and the time to failure were such that the customer was able to plan for a scheduled shutdown some three months out, as opposed to an unscheduled shutdown immediately. Production immediately went up to full rates, and condition

(continued)

> ### Condition monitoring story *(cont'd)*
>
> monitoring to track the deterioration in the gearbox and bearings was put into place by the supplier partner's engineers. The customer's process operators were then trained in the basic skills of condition monitoring and diagnosis.
>
> The end result was a delighted customer with short- and long-term solutions to an immediate crisis, and a supplier who consolidated their position as a long-term strategic partner, generating a clear point of difference from the competition.

The four stages of relationship maintenance

The work of engineers and operations specialists falls into four maintenance categories:

- breakdown maintenance

- preventative maintenance

- predictive maintenance

- design-out maintenance.

This division of maintenance expertise also provides useful insights into relationship maintenance. It is in the areas of predictive and design-out (innovation-based) maintenance that real competitive advantage lies. Preventing non-conformance, or the non-delivery of agreed requirements, is a minimum expectation for staying in business today. Prevention will keep you in the competitive pack. Innovative predictive and design-out strategies will keep you ahead of the pack, turning best practice into next practice. Here is how each maintenance approach applies to relationships.

Breakdown maintenance

How many times have we seen relationships managed on the basis of the breakdown maintenance principle, focused on the short-term? Knowingly or unknowingly the relationship runs to a point of failure, breakdown or crisis. Examples involve people, commercial pricing, business systems and processes, quality and service. We have all heard the expressions 'management by crisis', or 'we manage from one crisis to another'. This is a world typically built around a firefighting culture and a world of reactive complaint resolution, rather than a proactive search for longer term opportunities or solutions. Not only is this a tough and costly way to run a business, it is unrewarding and frustrating for all parties involved.

Operations people will tell you that unscheduled, unplanned breakdown maintenance on critical equipment can be the stuff of nightmares. There is no such thing as a pleasant surprise in the case of unplanned maintenance. It is a short-term approach and ultimately an unprofitable way to do business.

While a run-to-failure strategy may be appropriate for certain plant and equipment, it is rarely sustainable within the context of relationship management.

Preventative maintenance

The next alternative is prevention rather than cure. It sounds simple, and it is. Prevention in relationship management is about a total quality approach, and it is as much an attitude as a physical reality and business strategy. It is a mindset with a medium to longer term focus that asks: What can I do to ensure that certain things don't happen? For example, conducting regular business review and development meetings to prevent or avoid any potential people, process, product or technology issues. Co-location of customer and supplier employees, or internal operating groups and functions will encourage better coordination and information sharing, and stave off any relationship or communication breakdowns.

Predictive maintenance

The ability to predict breakdowns or problems allows for a managed long-term maintenance program to be effectively planned and implemented.

In business relationships, this is the world of real innovation and creativity. In this world, people at all levels are thinking about the next steps, adding value, and predicting and implementing tomorrow's initiative today. This is the first step beyond total quality, a proactive approach to thinking outside the square. Skill and leadership are critical here, as this is life in the fast lane, with a genuine long-term perspective.

For example, getting paid on the performance and effect of products and services (outcomes) and not on unit price (inputs) enables a different customer-supplier conversation and delivers sustainable competitive advantage.

Design-out maintenance

This is where Partnering and Alliancing, Pioneering and Community relationships deliver their rewards. One example of this is the one-team approach, with its seamless and transparent boundaries and integration of customer and supplier partners' personnel, systems and processes. Customer personnel, blue collar and white collar, actually report to supplier managers. Performance, measurement, attitudes and the sharing of risk and reward are managed openly and transparently between the partners for mutual benefit.

So why do relationships fail?

With business relationships being so important, why do so many fall apart or at least fail to achieve their full potential? It is suggested that less than 50 per cent of partnerships and alliances are successful. Nine out of 10 mergers and acquisitions fail to achieve expectations.

As it is with personal relationships, so it is in business. Almost without exception the problems, and ultimately failure, in business relationships can be traced back to people. Their basic beliefs, perceptions, decisions, actions or intentions will inevitably underlie a failed relationship. The manifestation of this relationship breakdown may be failure of plant and machinery, or of systems and networks, poor IFOT delivery performance, poor quality and financial performance, poor communication, diminished competitive advantage or reduced trust. But the root cause will rest with people. Some of the causes of relationship failure are:

- a change in senior management or CEO giving new direction at odds with the existing relationship strategy

- poor internal alignment on culture, strategy, structure, process and people

- decisions and judgments made for short-term financial gain rather than long-term objectives

- people at the operating levels, who must make the relationship work, not being informed about, or involved in, the relationship improvement process

- a lack of clarity in the partner's values, motivations and expectations

- lack of succession planning around key players — lose your champions, lose your will, then lose your way

- failure of senior management to lead and inspire the change process and continuous improvement journey

- failure to effectively manage or eliminate roadblocks or gatekeepers

- over-promising and under-delivering on agreed requirements and expectations

- permanent loss of trust

- the wrong level of investment in time and people.

It is the obligation of all those involved in the relationship to pick up and successfully manage all the stresses and strains that are involved. These are skills to be learnt, developed and coached, but they may mean the difference between success and failure. At the very least the effective application of relationship management skills will save much time, frustration and valuable resources.

Summary

- The relationship development curve underpins the concept of journey management in 0 to 10 Relationship Management.

- The five stages on the relationship development curve are:
  - instability
  - crisis
  - trust
  - innovation
  - no turning back.

- As with life generally, business relationships rarely run in a straight line. There are highlights and lowlights, points of crisis, periods of rapid change and periods where progress plateaus. Knowing where you are on the relationship development curve allows for effective course setting to the desired future state.

- Principle 5, Insanity is doing the same things and expecting different results, aligns directly to the relationship development curve.

- Smart organisations pre-empt and predict an impending crisis by honing their skills in crisis recognition and solution selling.

- Drop-offs from the relationship development curve can occur at any point because of complacency, changes in key people and relationship performance.

- Current paradigms form the platform for discovering new paradigms. This will be the basis for continuous and breakthrough improvement.

- Organisations need to regularly update the answer(s) to the value question, that is: what value is this relationship delivering for my organisation over the alternatives? Effectively answering the value question will provide a significant point of difference and competitive advantage for your organisation.

- It's essential to turn every problem into a solution, every crisis into an opportunity and every opportunity into reality.

CHAPTER 7
THE 12/12/6 ROADMAP PROCESS

While the 12/12/6 roadmap theme sits at the bottom edge of the simplified 0 to 10RM storyboard (figure 1.1 or the right-hand side of the full storyboard on the 0 to 10RM website at <www.0to10rm.com>) it should not be regarded as the end point of a process but rather another milestone on the journey of improvement. The 12/12/6 roadmap is often used to detail the 'What to do — tactical' level action plans associated with the Let's Go change model.

The journey of a thousand miles starts from beneath your feet.

Lao-Tzu[1]

The 12 motivators, the 12 process steps and the six outcomes provide the framework on which customer and supplier parties are chosen and the relationship developed, managed and improved. Motivators, process steps and outcomes combine to form the 12/12/6 roadmap (see figure 7.1, overleaf). The 12/12/6 process roadmap has three operating rules:

1   The steps are not numbered and they do not have to be completed in any particular order. Start anywhere. The choice or sequence of steps to be followed is infinitely flexible. The starting point is based on the requirements of the relationship parties and the operating environment, and where these parties are on the relationship improvement journey.
2   More than one step can be undertaken at any one time.
3   The process is continuous for the duration of the relationship.

Figure 7.1: the 12/12/6 roadmap

The 12 motivators of the roadmap

The 12 motivators form the driving forces behind the process. They provide the reasons and the motivations to propel the relationship forward. They are the drivers as to why people think, behave and do the things they do. The motivators complete the question of whether the next step will:

- add value

- reduce total costs

- improve communication

- develop trust

- resolve conflicts

- remove hidden agendas

- provide leadership

- empower people

- gain commitment

- develop ownership

- break down barriers

- remove fear.

Remember, the motivators are not listed in priority order.

Add value

Adding value involves the products, services, skills, technologies and other attributes that provide value and opportunities, unique or differentiated, that are of benefit to buyers and sellers beyond a low cost or cheap price. Adding value is also the basis for the value propositions that underpin the relationship(s).

Are the benefits achieved in the performance or effect of the product or service greater than the cost of creating them for the seller and greater than the premium paid by the buyer? Will more of the product or service be sold at the same price, or will other benefits, such as higher customer loyalty, referred business or longer term commitments, be achieved? Value adding might also involve the provision of products and services beyond the original products and services for which the supplier is being paid or the exploitation of synergies between the relationship parties. Are the benefits sustainable?

Reduce total costs

For buyer or seller, do the activities performed produce a lower total cost in the process, product or service? Total costs typically involve a combination of fixed, variable and opportunity costs.

Improve communication

Improved communication includes any interaction that helps individuals, groups and teams — internal or external — to relate to each other more effectively, irrespective of their status or role in the organisation; to have a clearer understanding of what other individuals, groups and teams do, and their roles, responsibilities, accountabilities; and to participate in those activities that will add value for the business. Sharing information in an open, honest, accurate, relevant and timely manner; maintaining a helpful, open attitude; and respecting and trusting others are the keys to good communication. Communication comes in many forms, verbal and non-verbal, face to face, and long distance through any or all of the available technologies.

Develop trust

Trust is a function of competence as well as character. It is about doing what you said you would do. Trust is based on integrity, respect, openness, honesty and clear communication. A central question is whether trust is earned or given. The measure of trust is how much you are prepared to rely on others without questioning, requiring evidence or examination. Trust can be placed in people, systems and processes, and presents as the ability to reliably deliver expected results.

Resolve conflicts

Conflicts often arise because of poor communication, misalignment, misunderstanding, non-conformance to agreed requirements or unmet expectations. Conflicts could result from an adversarial and combative approach to the relationship. Irrespective of whether conflicts arise by design or by accident, ultimately, they must be resolved if effective progress is to be achieved and sustained. The resolution process might include problem solving, mediation or, as a last resort, formal conflict resolution.

Remove hidden agendas

Hidden agendas are the issues, views, opinions or intentions that individuals or groups deliberately wish to conceal and which differ from their publicly held views or statements. Hidden agendas are typically full of hypocrisy and double standards. Removing them is a cornerstone for improved communication, in particular, and for supporting the other motivators. Unfortunately, hidden agendas can exist at any level within the organisation. Seek them out and remove them. They are destructive and debilitating.

Provide leadership

How can the relationship and associated business activities give genuine leaders the chance to lead and potential leaders the opportunity to learn? Leadership styles will be situational and vary according to need and circumstance. The relationship current state and desired future state, and the nature and extent of change required will influence the choice of leadership style.

Leadership is the art of accomplishing more than the science of management says is possible.

Colin Powell, former US secretary of state[2]

Empower people

Empowerment allows an individual or groups of individuals to acquire the knowledge, skills and attitudes they need to cope with changing life circumstances and a constantly changing world.

High levels of empowerment are directly related to employee engagement and wellbeing, which positively impacts productivity and overall business output. Empowerment is linked to clear roles, responsibilities and accountabilities. Empowerment improves initiative, judgement and motivation, leading to better quality decision making and more effective problem solving. Empowered people have a greater sense of self-belief, self-esteem and self-confidence. However, people should be empowered only as far as they are aligned with your vision, values and strategic goals. People can be engaged and enabled, but still not be empowered. There is nothing more destructive for an organisation, and specifically for relationship development, than smart people coming to work with a bad attitude.

The level of empowerment achieved will only be effective if the appropriate training is implemented and the necessary skills and competencies developed. With empowerment comes freedom.

Gain commitment

Making or gaining commitment from people in an organisation involves an implicit or explicit pledge or obligation. Commitments can be mutual or self-imposed. Commitment management is linked directly to trust and integrity. In short, doing what you said you would do. Commitments involve binding individuals, groups or organisations to a specific course of action for which they are directly accountable. A commitment must be realistic and achievable. Over-promising and under-delivering creates a potential playground for broken commitments.

Accountabilities and levels of authority need to be understood and communicated before commitments are made. Making a commitment at a personal level is the moral equivalent of giving your word or making a promise. Honour, integrity and credibility are at stake if commitments are not honoured. Broken commitments have a downstream ripple effect on the commitments made by others. Honoured commitments are essential to high performance relationship management, irrespective of the relationship type.

Develop ownership

Ownership looks like people and teams successfully handling responsibilities, tasks and accountabilities as if they were of their own design or creation. Ownership begins with giving employees recognition for delivering successful outcomes and for the personal commitment involved in achieving them. This involves empowering those individuals and teams with clear roles and purpose.

Break down barriers

Departmental, functional, operational or cultural barriers are just some of the road-blocks to effective communication and good business outcomes within and between organisations. Barriers might be real or perceived, physical or non-physical. They may be historical, cultural, personal, technology based or organisational. Barriers can be destructive and debilitating, and lead to the formation of organisational silos, factions and fiefdoms. Barriers create a legacy of higher costs, duplication, lower productivity and efficiency, poor communication, reduced effectiveness and increasingly frustrated employees.

Sometimes barriers are necessary—for example, in the defence and national security sectors; in classified research and development projects; where there is a need for confidentiality surrounding mergers and acquisitions, or defending and protecting tender or bid information that is market sensitive or commercial in confidence. If barriers are appropriate, effective communication and other motivators need to be in place to avoid or counter the negative downsides.

Remove fear

Edwards Deming, one of the founders of the Total Quality Management movement, wisely pronounced that 'Fear takes a horrible toll. Fear is all around, robbing people of their pride, hurting them, robbing them of a chance to contribute to the company. Replace fear with freedom and security.'[3]

Fear manifests itself in many forms: fear of asking questions; making mistakes, challenging the system; unemployment; failing to meet management or budget targets; the appraisal and performance system; new technology; and uncertainty about the future. Fear can be rational or irrational.

John McConnell, a Deming devotee, talks about fear in his book, *Safer than a known way*: 'Fear is negative and destructive. It destroys morale and teamwork, it corrupts data; it damages quality and thus productivity. I do not believe any management theory can work in a company driven by fear, let alone a theory that requires fundamental change in attitudes and methods'.[4] An intuitive leader will neutralise fear when it arises.

Type 1 Combative and Type 2 Tribal relationships are the only relationships on the 0 to 10RM scale that may in any way gain benefits from instilling fear into employees, customers or suppliers. Even then the outcome is rarely anything other than a short-term win–lose, one-sided result.

The 12 steps of the roadmap

The 12 process steps provide the vehicle and structure by which the 12 motivators, as the drivers of the process, can be realised. The 12 process steps are:

- Evaluate and select the vendor, supplier or partner.
- Review internal relationships for quality and performance.
- Review or workshop the relationship process, progress and performance.
- Complete requirements analysis for the present and future.
- Meet or exceed requirements IFOTA1 specification, for quick wins or long-term solutions.
- Select or review relationship managers and teams.
- Conduct site visits.
- Review or implement skills requirements.
- Review supplier and support relationships upstream.
- Establish technology requirements, current and future.
- Review internal and external relationship networks.
- Develop, implement and review relationship strategies and action plans.

Like the 12 motivators, the 12 steps are not numbered because there is no order of priority and the journey can begin at any point. In the following checklists, identify specific actions to implement for each step according to the needs of the organisation and the state of the current relationships.

Evaluate and select the vendor, supplier or partner

Objective: To evaluate and select the right customer or supplier for the right reasons with the right relationship approach (see figure 7.2, overleaf).

This step is predicated on understanding the relationship types that need to be developed, and the value propositions and key objectives that need to be delivered to ensure the business or organisational strategies are achieved. Evaluation techniques range from compiling and checking off simple checklists to the use of complex models.

The choice will depend on the operating environment, market sector, nature and size of the relationship(s) to be developed, the products or services and organisations(s) or people involved. The relationship alignment diagnostic (RAD) and relationship strategy maps (RSMs) will prove particularly helpful in evaluation and selection. This step is equally applicable to suppliers and customers.

Figure 7.2: evaluating and selecting a vendor, supplier or partner

Figure 7.3 is a worked example to demonstrate the connection between strategic value and commercial value, willingness and capability discussed in chapter 2 to the evaluation and selection process.

Checklist of key points and action plans to evaluate and select vendor, supplier or partner

- Agree on the scope of work activities, products and services involved, and document the associated value propositions (that is, those benefits or opportunities beyond a low cost or cheap price) that support relationship engagement.

Figure 7.3: sample of a partner evaluation worksheet

Definitions

%WA = percentage weightings for value statements
%WB = percentage weightings for delivery statements
%WC = percentage weightings for evaluation focus
%TW = %WA x %WB x %WC = Total Weighting AxBxC

0 to 10RM performance scoring scale

1 = Unsustainable	6 = Good
2 = Poor	7 = Excellent
3 = Below average	8 = Outstanding
4 = Fair	9 = World class
5 = Satisfactory	10 = Superior

A. Value Statement	%WA	B. Delivery Statement	%WB	C. Evaluation Focus	%WC	%TW	Company X Raw Score	Company X Weighted Score	Company Y Raw Score	Company Y Weighted Score	Company Z Raw Score	Company Z Weighted Score
1. Improve safety, health and environmental performance.	15	1.1 Effective safety, health and environmental systems and management practices in place.	40	1.1.1 Strategic value	30	1.8	9	0.16	8	0.14	9	0.16
				1.1.2 Commercial value	25	1.5	8	0.12	7	0.11	8	0.12
				1.1.3 Willingness	15	0.9	7	0.06	5	0.05	7	0.06
				1.1.4 Capability	30	1.8	8	0.14	6	0.11	6.5	0.12
				Total	100	6						
		1.2 Evidence of a strong commitment to and outstanding performance in providing a safe and healthy workplace.	30	1.2.1 Strategic value	25	1.1	9	0.10	8	0.09	9	0.10
				1.2.2 Commercial value	25	1.1	8	0.09	7	0.08	8	0.09
				1.2.3 Willingness	25	1.1	7	0.08	5	0.06	6	0.07
				1.2.4 Capability	25	1.1	8	0.09	7	0.08	8	0.09
				Total	100	4.5						
		1.3 Demonstrated commitment to and strong achievements in protecting the environment.	30	1.3.1 Strategic value	25	1.1	9	0.10	5	0.06	6	0.07
				1.3.2 Commercial value	25	1.1	8	0.09	7	0.08	8	0.09
				1.3.3 Willingness	25	1.1	7	0.08	6	0.07	7	0.08
				1.3.4 Capability	25	1.1	7	0.08	6	0.07	7	0.08
				Total	100	4.5						
				Total - Safety, Health, Environment	100	15		1.20		0.98		1.13
2. Deliver sustainable and superior value which supports the broader business strategy.	30	2.1 Demonstrated understanding of customer requirements and strategy expectations.	20	2.1.1 Strategic value	25	1.5	7	0.11	6	0.09	7	0.11
				2.1.2 Commercial value	20	1.2	8	0.10	4	0.05	5	0.06
				2.1.3 Willingness	25	1.5	7	0.11	4	0.06	4	0.06
				2.1.4 Capability	30	1.8	8	0.14	5	0.09	4	0.07
				Total	100	6						
		2.2 Credible plans or milestones for transition phase, ongoing product or service development and delivery, operations and maintenance.	10	2.2.1 Strategic value	30	0.9	8	0.07	5	0.05	5	0.05
				2.2.2 Commercial value	25	0.8	9	0.07	5	0.04	5	0.04
				2.2.3 Willingness	20	0.6	8	0.05	9	0.05	9	0.05
				2.2.4 Capability	25	0.8	7	0.05	5	0.04	5	0.04
				Total	100	3						
		2.3 Demonstrable track record of delivering best value for money performance.	10	2.3.1 Strategic value	25	0.8	9	0.07	5	0.04	6	0.05
				2.3.2 Commercial value	25	0.8	8	0.06	6	0.05	6	0.05
				2.3.3 Willingness	20	0.6	8	0.05	7	0.04	8	0.05
				2.3.4 Capability	30	0.9	8	0.07	6	0.05	6	0.05
				Total	100	3						
		2.4 A commercial arrangement that ensures the best value for money or total cost outcomes.	50	2.4.1 Strategic value	20	3	8	0.24	5	0.15	5	0.15
				2.4.2 Commercial value	50	7.5	9	0.68	5	0.38	5	0.38
				2.4.3 Willingness	10	1.5	8	0.12	5	0.08	5	0.08
				2.4.4 Capability	20	3	8	0.24	5	0.15	5	0.15
				Total	100	15						

(continued)

Figure 7.3 (cont'd) : sample of a partner evaluation worksheet

Section	Criterion	Wt	Score	Val	C1	Score	C2	Score	C3
10 — 2.5 Strength of balance sheet	2.5.1 Strategic value	35	8	1.1	0.08	5	0.05	5	0.05
	2.5.2 Commercial value	35	8	1.1	0.08	6	0.06	5	0.05
	2.5.3 Willingness	10	8	0.3	0.02	5	0.02	5	0.02
	2.5.4 Capability	20	8	0.6	0.05	5	0.03	5	0.03
	Total	100		3					
	100 Total – Superior Value for Money				2.45		1.55		1.56
30 — 3. Establish a role model partnering relationship which is performance based, risk–reward linked, open, transparent, collaborative and mutual beneficial.	**25 — 3.1 Understanding of Partnering and Alliance principles and practice.**								
	3.1.1 Strategic value	20	9	1.5	0.14	5	0.08	6	0.09
	3.1.2 Commercial value	20	8	1.5	0.12	6	0.09	5	0.08
	3.1.3 Willingness	30	9	2.3	0.20	5	0.14	7	0.16
	3.1.4 Capability	30	8	2.3	0.18	4.5	0.10	4.5	0.10
	Total	100		7.5					
	20 — 3.2 Demonstrated Partnering and Alliancing capability.								
	3.2.1 Strategic value	20	8	1.2	0.10	6	0.07	6.5	0.08
	3.2.2 Commercial value	20	8	1.2	0.10	6	0.07	5	0.06
	3.2.3 Willingness	30	9	1.8	0.16	6	0.11	7	0.13
	3.2.4 Capability	30	8	1.8	0.14	4	0.07	4.5	0.08
	Total	100		6					
	30 — 3.3 Commitment to working in a performance based, risk–reward linked, open, transparent, collaborative, mutually beneficial relationship.								
	3.3.1 Strategic value	20	9	1.8	0.16	5	0.09	6	0.11
	3.3.2 Commercial value	20	8	1.8	0.14	6	0.11	5	0.09
	3.3.3 Willingness	30	8	2.7	0.22	6	0.16	7	0.19
	3.3.4 Capability	30	9	2.7	0.24	4.5	0.12	7	0.19
	Total	100		9					
	25 — 3.4 Ability to work collaboratively with and manage third parties.								
	3.4.1 Strategic value	20	8	1.5	0.12	6	0.09	6	0.09
	3.4.2 Commercial value	20	8	1.5	0.12	5	0.08	6	0.09
	3.4.3 Willingness	30	9	2.3	0.20	7	0.16	7	0.16
	3.4.4 Capability	30	9	2.3	0.20	4.5	0.10	7	0.16
	Total	100		7.5					
	100 Total – Partnering and Alliancing				2.55		1.63		1.84
25 — 4. Deliver world class or best practice innovation through continuous and breakthrough improvement.	**60 — 4.1 Demonstrated track record of delivering successful innovation and improvement.**								
	4.1.1 Strategic value	20	8	3	0.24	7	0.21	8	0.24
	4.1.2 Commercial value	20	8	3	0.24	8	0.24	8	0.24
	4.1.3 Willingness	20	10	3	0.30	9	0.27	10	0.30
	4.1.4 Capability	40	9	6	0.54	5	0.30	7	0.42
	Total	100		15					
	40 — 4.2 Commitment to leverage partners' global capability and fully engage partner resources in the development and delivery of products and services.								
	4.2.1 Strategic value	50	8	3	0.24	7	0.21	8	0.24
	4.2.2 Commercial value	50	8	5	0.40	8	0.40	8	0.40
	4.2.3 Willingness	10	8	1	0.08	7	0.07	8	0.08
	4.2.4 Capability	10	9	1	0.09	6	0.06	7	0.07
	Total	100		10					
	100 Total – Innovation				2.13		1.76		1.99
100 TOTAL					8.33		5.92		6.52

Value statements 1 to 4

- Agree on the desired future state for the relationship; that is, the relationship approach and performance levels. Follow these steps in any order as appropriate to the situation in your organisation:
  - analyse the strategic value and commercial value of the products and services involved
  - establish willingness and capability of the potential relationship candidates
  - use the eRAD (see chapter 2) as a pre-work exercise to frame this analysis or as a follow-up analysis (or both) to monitor the relationship status
  - use relationship strategy maps as appropriate.
- Develop the relationship engagement strategy and confirm alignment of agreed scope, value propositions and delivery process with the broader corporate or organisation strategy or strategies.
- Agree with key stakeholders if this is a go to market process or negotiated outcomes process.
- Complete a SWOT (strengths, weaknesses, opportunities, threats) and risk analysis for the relationship strategy and evaluation or selection process to be deployed.
- Review the impact that the relationship will have on existing and future markets, and customer and supplier relationships.
- Understand and map the internal value chain and the external supply chain to understand the degree of alignment.
- If appropriate, agree on the role of third-party professionals, staff subject matter experts, lawyers, consultants and deal-brokers within the process.
- Evaluate and select the vendor, supplier or partner based on the strategic value and commercial value they bring to the value propositions; and their willingness and capability to meet or exceed relationship scope requirements.
- Identify supporting vendor, supplier or partner relationships (internal and external) and review the requirements and desired performance levels.
- Complete the selection process (see figure 7.4, overleaf) through the use of a cross-sectional team possessing the right level of skills, functional engagement, and strategic influence with shared understanding. Some of these team members will or should be involved in the transition and implementation stages.
- Conduct desk top evaluation, internal and external workshops appropriate to the desired relationship type to be engaged.
- Understand competitor strategies and develop contingency plans.

Figure 7.4: typical steps in the evaluation or selection process for Partnering and Alliancing relationships

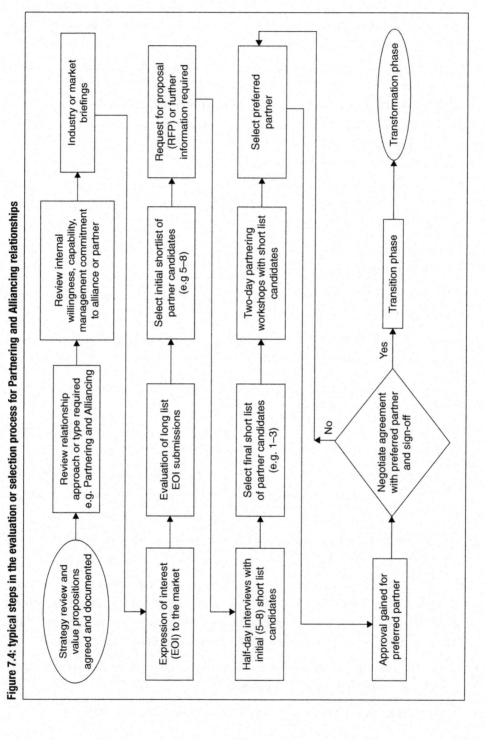

Figure 7.4 shows the common selection process steps for a strategic Partnering and Alliancing relationship engagement, which can be adapted to the internal and external operating environment.

Review internal relationships for quality and performance

Objective: To build and sustain internal willingness and capability for high performance relationship management.

As discussed in chapter 6, one of the biggest impediments to the effective delivery of external strategy is the quality of internal relationships (see figure 7.5). In other words know thyself. Strong leaders in good organisations unite employees around a common purpose. Doing so at the shopfloor and mid levels of the organisation helps to break down internal barriers and improve employee engagement and productivity.

Figure 7.5: review internal relationships for quality and performance

Checklist of key points and actions to review internal relationships for quality and performance

- Complete internal communication and relationship maps, and implement action plans for improvement using an understanding of your organisation's primary and support value chain activities.

- Conduct an internal eRAD relationship health check (see chapter 2) between selected departments, functions and business units.

- Analyse the quality of communications within the organisation. Agree and implement open, honest, accurate, transparent, timely, relevant communication and information sharing.

- Set in place the goal to earn and give trust; to document what trust looks like, for example, in attitudes, behaviours and practices (trust charter; see chapter 4).

- Focus on sharing information, engaging and enabling, empowering, gaining commitment, and developing ownership around common goals.

- Take action:
  - communicate, listen and deliver on commitments
  - resolve conflicts between functions and departments
  - clarify roles
  - expose hidden agendas
  - develop internal willingness and capability.

- Implement an internal partnering strategy between business units, departments and functions as appropriate.

- Conduct an analysis of the organisation's (internal) willingness and capability to partner.

- Develop an understanding around high performance relationship principles and practices.

- Agree and document individual and team roles and responsibilities associated with high performance relationships.

- Identify and develop plans to support internal innovators, change and relationship champions, key influencers for relationship improvement, such as relationship facilitators and relationship managers (see chapter 5).

- Conduct facilitated workshops in support of the innovation and improvement process, such as relationship review and improvement workshops, value engineering and innovation workshops, process review and re-engineering workshops, technology review workshops, problem solving workshops, and requirements review and reset workshops.

Review or workshop the relationship process, progress and performance

Objective: To review or workshop the relationship process, progress and performance.

Relationship review workshops are one of the main mechanisms for reviewing relationship process, progress and performance (see figure 7.6). The nature and structure of the review workshop will reflect the 0 to 10 relationship type. This could involve a two-hour meeting or a two-day facilitated offsite workshop.

Figure 7.6: conduct reviews or workshop the relationship process progress and performance

For Type 6 Major and in particular Type 7 Key relationships, business review and development meetings (BRADs) should be held regularly; for example, half-yearly to annually. Type 8 Partnering and Alliancing, Type 9 Pioneering and Type 10 Community relationship reviews are typically held quarterly, half-yearly or annually, and take the form of joint one- to two-day facilitated workshops (appendix C has a sample Relationship review and improvement workshop agenda). Vendor relationships (Type 1 Combative, Type 2 Tribal, Type 3 Trading and Type 4 Transactional) typically require only one-sided reviews, conducted in-house for internal consumption. There are exceptions, but they tend to be irregular and driven by specific events rather than forming part of a structured process review. Type 5 Basic relationship reviews typically take place somewhere between the *do* and the *charge* to review and sign-off on work completed. Depending on the scope and complexity of the work involved this can be a simple or complex process.

Checklist of key points and actions to review or workshop the relationship process, progress and performance

- Agree on the background, purpose and desired outcomes from the relationship review or workshop; for example, to review the current state and desired future state for the relationship and jointly agree on a plan to bridge the gap in direct support of agreed business objectives.

- Agree on the structure, format, style, such as formal or informal; structured or unstructured agenda and outcomes reporting; one-party, two-party or multi-party involvement.

- Agree on the workshop or meeting participants list based on those people who best able to contribute to the delivery of the purpose and desired outcomes; for example, sponsor or chair, facilitator, presenter, subject matter experts, participants, observer(s) and minute taker.

- Agree on detailed agenda, including agenda items; who will lead; and the process tools and techniques to be used (data presentation, participant interaction, breakout groups; time allocated). See appendix C for a draft one-day relationship review and improvement workshop agenda as a guide.

- Ensure an effective follow-up process and ownership of assigned actions.

- Agree on the time, place and any non-standard agenda items for next review meeting or workshop.

- Outside of these high level review and improvement workshops, conduct other engagement, subject-specific workshops; work team or project workshops; toolbox meetings and informal discussions at all levels between the relationship parties to share information and build trust, report on progress (performance against objectives and external and internal benchmarks) to wider stakeholder groups; to review new opportunities for innovation, continuous and break-through improvement, and profitable growth; to challenge, inspire and lead the relationship improvement process; to have fun and celebrate success; and to induct, train or educate new relationship participants.

- Conduct formal and informal discussions on relationship performance and future direction with stakeholders and other key influencers.

- Use this review and workshop process, formally and informally, to ensure that the people at the operating levels who are charged with making the relationship work are informed, involved and committed to the relationship goals and objectives.

- Use relationship alignment diagnostics (off-line and online; for example, using the eRAD) or relationship health checks (see chapter 2) on a regular basis to track and feedback progress and performance.

- Align the review and improvement process with agreed relationship governance structures and processes.

The purpose and benefits of workshops

When done well, for the right reasons with the right people involved, workshops for relationship review and improvement can be watershed events that produce breakthrough outcomes. They can deliver outcomes, commitments and a momentum that would otherwise not be possible. On the relationship development curve there are numerous touch points where different workshop types can be immensely helpful (see figure 7.7, overleaf).

A well-facilitated workshop provides the opportunity and environment, as well as a process for groups or teams to work together (at the one place and time), to achieve clear business goals and objectives. The purpose of facilitated workshops is to achieve the agreed workshop goals and business objectives through:

- focused pre-workshop preparation

- effective workshop facilitation

- planned post-workshop and follow-up activities.

The real success of any workshop will be demonstrated in the practices and attitudes displayed and the results achieved in the weeks and months and years after the workshop.

Figure 7.7: the relationship development curve overlaid with types of workshops

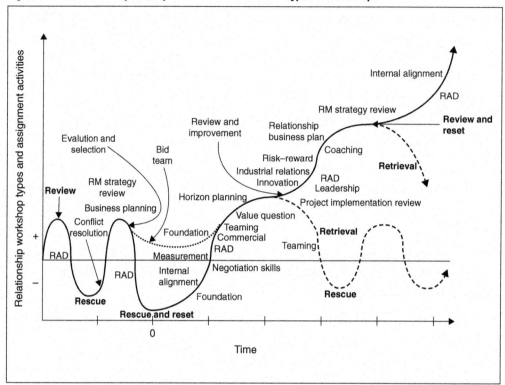

Conduct requirements analysis for the present and future

Objective: To understand and mutually agree on requirements for the present and the future.

Seek first to understand, then be understood.

Stephen R. Covey[5]

Requirements are directly linked to open, honest, timely and accurate information sharing and communication. They are linked to the value propositions, purpose or intent for the relationship. An understanding of requirements is accelerated if it is based on a platform of trust. In high performance relationships, current and future requirements of customers and suppliers need to be unambiguous, documented and shared with those people charged with their delivery (see figure 7.8).

Figure 7.8: conduct requirements analysis for the present and future

Understanding and meeting customer requirements IFOTA1 specification is the starting point for all differentiation and innovation. Sometimes called getting the basics right, this involves a two-way flow of ideas, data and other information, a clear understanding of the processes involved, and the willingness and capability of the parties to deliver. Requirements are two-way, each party having both responsibilities and commitments. For example, suppliers need to deliver products and services IFOTA1, just as customers need to give reasonable lead times and pay bills IFOT. Put simply, all parties involved need to do what they say they will do and act in good faith.

Checklist of key points and actions to conduct requirements analysis for the present and future

- Start with a skilful and probing use of open and closed questions based on good preparation and background knowledge of the operating environment. Be positive and treat the uncovering of requirements as a journey of discovery.

If necessary, use requirements teams, or subject matter experts, or call in third-party professionals for complex relationships. Requirements are two way, so expect or initiate return questions to gain full understanding and to build trust.

- Follow up with active listening and a documented capture of mutually understood and agreed customer needs and requirements. Listen (to understand); question (to understand); and confirm (to ensure common understanding).

- Turn product, service or systems and process features into benefits.

- To build trust and trustworthiness, reinforce your organisation's willingness and capability to meet and, if appropriate, to exceed requirements. Be truthful and authentic.

- Use clear and open communication to manage concerns. Deal with sceptism by offering proof; misunderstanding by reinforcing the link between features and benefits and negative perceptions or indifference by focusing on the wider picture, probing for unrealised needs and reinforcing previously accepted benefits.

- Only engage in negotiations when the willingness to buy and sell are equal and always trade concessions around core and related issues rather than giving them away.

- In line with the relationship approach engaged, negotiate and document scope, prices and costs, terms and conditions for delivering the requirements and as appropriate associated contracts, agreements and service level agreements.

- Once customer and supplier partner requirements (current and future) are jointly identified, document and share them with all those involved in their delivery.

- Analyse work process flows, internal value chains and external supply chains to assess opportunities for improvement and re-engineering.

- Review agreed requirements regularly and benchmark against agreed baselines.

- Link requirements to metrics (key performance indicators, or KPIs) to measure and manage performance and drive innovation and improvement.

- Uncover and solve problems that are not of your making, and use the process and outcomes as a point of difference.

- Repeat any of these steps if gaps in understanding or information are discovered.

Meet or exceed requirements IFOTA1 specification

Objective: To meet or exceed requirements IFOTA1 specification.

Meeting or exceeding requirements is about implementing quick wins or quick fixes, as well as implementing longer term solutions (see figure 7.9). The quick fix is a temporary solution until the problem recurs or until permanent, quality solutions have been found. While a clever and effective quick fix may be the start of a high performance relationship, it will in no way sustain it. The future will be about flexibility, consistency, and the reliability and dependability of products, services, systems and processes as a base upon which the relationship can develop.

Figure 7.9: meet or exceed requirements to IFOTA1 specification

Checklist of key points and actions to meet or exceed requirements to IFOTA1 specification for quick wins, quick fixes or temporary solutions

- Ensure that current performance levels against jointly agreed requirements are known. Measure to determine degree of conformance.

- Implement quick fixes; identify and take advantage of short-term opportunities; and capture the low-hanging fruit quick wins.

- Search for problems and crises that can be turned into opportunities quickly and cost effectively.

- Lay out the foundations for permanent solutions, for example through implementing systems, processes, procedures and ensure process stability.

- Develop and implement a 100-day improvement plan (a Horizon 1 Plan).

Checklist of key points and actions to meet or exceed requirements to IFOTA1 specifications for quality longer term solutions

- Develop and implement medium to longer term improvement plans as and when required.

- Ensure quality and other systems and communication protocols are in place to meet or exceed current and future requirements.

- Maintain open, honest, timely, accurate, continuous communications and information sharing around requirements, such as performance to date, trends and upcoming changes.

- Ensure processes in general, such as procurement, materials handling, planning and design, research and development, production, and marketing and sales, are capable of consistently delivering internal and external customer or supplier requirements. Look for business improvement, value adding, business re-engineering and process engineering opportunities.

- Develop systems and processes based on the prevention, prediction and designing-out of special causes of non-conformances and unwanted variation to implement quality, fit for purpose solutions, and continuous and break-through improvement.

- Ensure internal and external relationships are fit for purpose to effectively deliver current and future requirements.

- Put effective incentive and recognition programs in place.

- Identify opportunities for achieving broader business outcomes and profitable growth. Action and review regularly.

- Conduct regular reviews of performance and feedback against requirements.

- Review the relevance of the other 11 steps to ensure that requirements are being met or exceeded.

Select or review relationship managers and teams

Objective: To ensure the right people, in particular relationship managers and support teams, are in place to deliver the best possible performance outcomes.

A team is a small number of people with complementary skills who are committed to a common purpose, set of performance goals and approach for which they hold themselves mutually accountable.

Jon Katenback and Douglas Smith[6]

Relationship managers play a critical role in the relationship management and improvement process (see chapter 5 for more information). The relationship manager (see figure 7.10) possesses the clarity of purpose and the right skill sets, and has the networks and contacts to take ownership of the health and wellbeing of the relationship.

Figure 7.10: select and review relationship managers and support teams

The impact of the relationship manager is compounded by the high performance teams and groups they work with. A typical governance structure in support of a Type 8 Partnering and Alliancing relationship is shown in figure 7.11.

Checklist of key points and actions to select or review relationship managers and teams

- Have in place high performance relationship managers who are fit for purpose, and have the knowledge, skills and attributes (competencies) that align with the overall requirements and value propositions underpinning the relationship.

- Ensure relationship managers have clear and documented roles and responsibilities, with associated performance metrics in place (see appendix B).

- As appropriate to the operating environment, relationship type and performance levels, engage support teams, wider integrated project teams, special purpose teams and subject matter experts to support the relationship manager in meeting or exceeding requirements.

- Implement the following 10 elements[7] to ensure teams are effective:
  - clarity of team goals
  - an improvement plan (to deliver goals)
  - clearly defined roles
  - clear communication
  - agreed and beneficial team behaviours (skills and practices)
  - well-defined decision-making processes
  - balanced participation
  - agreed and established team ground rules (norms)
  - an awareness of the group process (teaming)
  - use of the scientific or quality approach (data-centric).

- Establish joint leadership and management teams with clear roles and objectives, and accountable KPIs for complex partner segment relationships (see figure 7.11).

- Select team members on the basis of best person for the job, irrespective of the organisation they work for.

- Regularly review relationship manager and team(s) performance against agreed objectives, metrics and KPIs, and put in place appropriate incentive and recognition programs.

- Ensure that appropriate empowerment and authority levels are given.

Figure 7.11: typical governance structure for a Type 8 Partnering and Alliancing relationship

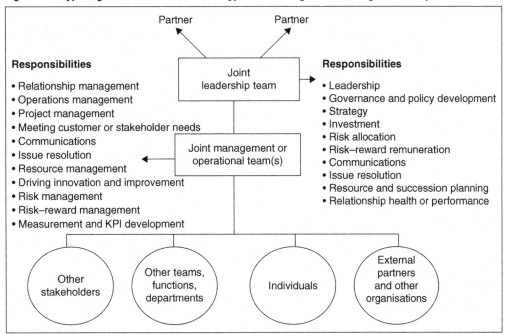

- Put team sponsors, executive oversight or reporting or governance structures in place.

- Schedule regular meetings for relationship review and improvement to enhance communication.

- Put in place succession plans for key team members and other key influencers.

- Establish and ensure that links with other teams, stakeholders, individuals, functions, departments, business units and selected third parties are identified and working effectively.

Conduct site visits

Objective: To bring people together, see and learn new things, share and benchmark information and ideas, build trust and develop friendships.

I was doing some consulting work on a very large mine site that had associated production facilities. Many hardened organisational silos, with defend and protect attitudes, behaviours and practices had developed over decades. Overall production output, asset reliability and productivity were being severely affected by poor communication, misaligned objectives and performance metrics and behaviours that were fundamentally based on self-interest.

In a conversation one day with a mine production operator of more than 25 years, experience I asked him when was the last time he had met the people and visited the downstream treatment plant, which was the next in line production facility. His answer staggered me. Never! In 25 years he had never visited the recipients of his work and hard labour. And after 25 years in his silo he had lost the ability or willingness to care.

Within weeks of a series of two-way site visits by the key players at both facilities, all the performance indicators started to improve and the attitudes, commitment and ownership of individuals had fundamentally changed for the better.

It is a lot easier to disappoint people you have never met and know little about. It's a lot harder to disappoint a friend who has trust in you and your organisation, function or department. Site visits, internal or external, are special events that can lead to relationship revitalisation (see figure 7.12). Don't underestimate their potential to restart or turn around a relationship for the better. Great things happen from bringing people together.

Figure 7.12: conduct site visits

Joel Barker[8] says the person most likely to change your paradigm is an outsider, someone who has little or no vested interest in your current paradigm. Two-way site visits will prove invaluable as catalysts for change and in developing and sustaining the relationship.

Checklist of key points and actions for conducting site visits

- Conduct site visits regularly to meet the people to bring them closer together. Engage the shopfloor as well as the top floor. Understand the operating culture — the way we do things around here. Communicate the importance of meeting or exceeding requirements. Build ownership and commitment at all levels. Share information and resolve issues. Discuss opportunities for improvement; innovate and cross-fertilise ideas. Build trust and develop friendships.

- Conduct third-party benchmark site visits to come to understand the opportunities for innovation and improvement.

- Establish and communicate a common sense set of rules and guidelines for site visits. For example: agree on primary objectives for the site visit; agree on dates, times, agenda, participant numbers; agree on roles, such as host sponsors, facilitators, team leaders, tour guides; agree on logistics, such as travel arrangements to and from each site, site inductions and refreshments; conduct pre–site tour briefings to review current relationship issues and opportunities; identify areas of special interest; agree ground rules and expected site visit behaviours, such as punctuality, politeness, appropriate dress, having an open attitude, being interactive and asking plenty of questions and respecting confidentialities; and conduct follow-up review discussions around idea implementation and the lessons learnt.

- Conduct internal and external site visits.

- Use high performance sites as role models, reference points and benchmarks for other parts of the organisation.

Review or implement skills requirements

Objective: To create a learning environment where skills and competencies are continuously developed to support and drive the ongoing innovation that sustains relationship improvement.

Investment in human and intellectual capital is one of the cornerstones to sustaining high performance business relationships. Good people in good organisations want to continually learn and contribute. This is not about throwing training courses at unsuspecting subordinates and ticking the boxes on completion. Combining an individual's existing knowledge, skills and personal attributes into competencies that are connected in some way to an organisation's vision and values is a most sustaining and powerful competitive force (see figure 7.13). When self-interest is aligned with mutual benefit the outcome is a more productive, empowered, self-confident and committed workforce. High performance relationships provide the added opportunity for shared learning, with a one-team focus around common goals. Combine all this with great products or services, and the results are obvious.

Figure 7.13: review or implement skills requirements

Checklist of key points and actions to review or implement skills requirements

- Determine the skills required to meet current and future customer requirements and review them regularly.

- Conduct or update a skills audit regularly.

- Identify skills gaps and develop an improvement plan.

- Implement personal development and training programs with ongoing review of competency levels, such as Partnering and Alliancing management; problem solving; conflict resolution; IT and PC skills; negotiation skills; professional selling and procurement skills; process mapping; business process re-engineering; high performance teaming; and leadership skills.

- Develop formal links to educational institutions as required.

- Identify opportunities for joint training between the relationship parties.

- Consider people exchange programs and secondments.

- Evaluate co-location opportunities for relationship participants and other selected third parties as appropriate.

- Build relationship management as a core competency to add value and create a point of difference for the organisation in the market place.

- Identify and develop internal relationship facilitators to support leaders and the delivery of business improvement goals. The relationship facilitators act as change champions, coaches, internal consultants, trainers, facilitators and provocateurs.

Review upstream supplier and support relationships

Objective: To improve upstream supplier and support relationships.

Develop a high performance relationship strategy both downstream with customers and upstream with suppliers, because the integrated supply chain is only as strong as its weakest link. The quality, availability and dependability of supplier raw materials, plant and equipment, spare parts, services, hardware, software, advice and information will all have a critical bearing on your ability to meet or exceed current and future customer requirements. While customers are the reason suppliers exist, you should not assume that supplier relationships are any less important. The quality of upstream relationships with suppliers, the quality of internal relationships between business units, and support functions will need to be consistent with the quality of the downstream relationships with customers (see figure 7.14, overleaf).

Ask the question: how much of your organisation's revenue is transferred to suppliers? The answer is probably somewhere between 50 per cent and 80 per cent. This reinforces the fact that high performing supplier relationships need to be a critical part of the business and relationship management strategy. Central to this will be building supplier capability and willingness based on trust, open communications and shared interests.

Figure 7.14: review upstream supplier and support relationships

Checklist of key points and actions for review supplier and support relationships upstream

- Undertake a wide-ranging supplier segmentation analysis. Identify important supplier relationships and review the current states and desired future states for the relationships using the 0 to 10RM relationship alignment diagnostic (RAD), eRAD (online relationship health check) and 0 to 10RM relationship strategy map tools (see chapter 2).

- Implement improvement plans to bridge the gap between the current state and desired future states. Jointly search for benefits based on agreed requirements.

- Evaluate or implement a supplier reduction program to integrate services, reduce variation and total cost of ownership, improve quality and performance.

- Have first tier suppliers manage second tier suppliers.

- Establish performance objectives, metrics, KPIs and stretch targets, and manage for improvement.

- Consider co-locating suppliers, both internally and externally.

- Encourage two-way training and shared learning to build trust, teamwork and innovation.

- Form joint supplier business improvement and project teams with clear purposes and focused goals.

- Develop performance-based arrangements and Partnering and Alliancing relationships with selected suppliers; namely, seamless interfaces and relationships, integrated systems, shared risk–reward aligned to common goals.

- Create a collaborative benchmarking process between suppliers to develop and share best practices for mutual benefit.

- Encourage and provide incentives for innovation, continuous improvement and breakthrough thinking with the top supplier group.

- Develop co-supplier relationships with competitors or product or service complementors to enhance competitive advantage and superior delivery performance to customers.

- Conduct regular reviews and town hall meetings with top suppliers for feedback and improvement.

- Conduct regular supplier best and next practice forums and seminars.

Establish technology requirements, current and future

Objective: To understand the opportunities for changes in technology and linking those changes to agreed requirements, both current and future.

All businesses or industry sectors today are technology intensive. The words of respected business management academics Gary Hamel and CK Prahalad in their book *Competing for the future* are as relevant today as they were when written in the 1990s.

Lacking a point of view about customers' future needs, there is a danger that a company will invest only in those technologies that correspond to current expressed needs. This is short-sighted. The link between technology and customers is not just currently articulated needs, but also product and service concepts that promise to satisfy unarticulated needs. The goal is to be neither narrowly technology-driven nor narrowly customer-driven. The goal is thus to be broadly benefits-driven, constantly searching for, investing in, and mastering the technology that will bring unarticulated benefits to humankind.

Gary Hamel and C.K. Prahalad[9]

Technology is a key enabler for higher purpose outcomes but, ironically, it is relationships that provide the step that determines the rate of delivery of those business benefits (see figure 7.15). The critical variables are trust, people, communications and attitudes. Thus relationships and technology are interdependent: one can't survive without the other.

Figure 7.15: establish technology requirements, current and future

Checklist of key points and actions to establish technology requirements, current and future

- Complete a technology gap analysis (current and future) in alignment with each relationship party's business strategy, the relationship value propositions (see chapter 4) and customer requirements (current and future). Do this throughout all business functions.

- Benchmark current technologies against alternative technologies and appropriate baseline data.

- Ensure that action plans or a technology roadmap is in place for improvement. Determine feasibility and fit with business strategy. Understand the timeframe and costs for implementation, development, ongoing maintenance and operation. Determine the nature of competitive advantage to be achieved through the application of the chosen technology. Evaluate the impact of its introduction on people and skills.

- Establish a process for regular review and benchmarking of technology requirements.

- Identify benefits and explore opportunities for technology integration and technology exchange. Identify internal areas of technology excellence that can be leveraged and exploited internally, between business units and functions, and externally, with customers and suppliers.

- Understand the impact of new technologies on strategy, development costs, competitors, resource requirements and people.

- Reach agreement with the relationship parties on the strategy and detail of any technology change, the impact on costs and benefits overall, and the role each party needs to play.

- Evaluate the role of acquisitions, joint ventures and licensing in the technology strategy, and implement actions as appropriate.

- Review the potential for technology integration, exchange or exploitation among business units, functions or departments, and with external customers and suppliers.

- Evaluate opportunities for process, procedures and systems alignment or integration that support the technologies and the associated relationship(s), and put improvement plans in place.

Review internal and external relationship networks

Objective: To understand the application and benefits derived from developing networks within and between the relationship parties and organisations.

Collaborative networks and communities, internal and external, are vehicles for supporting high performance relationships and organisational change (see figure 7.16). They range from the unofficial, self-organising, below the radar, independent pods and clusters, to formal, above-radar networks and communities integrated into the company's formal management structures. These formal networks have specific goals, clear roles, explicit accountability and executive oversight. The right people, in the right communities, for the right reasons, making good use of collaborative technologies and social media are a powerful and sustainable force for organisational change. When linked with special purpose project- and relationship-based work teams they are an ideal, cost-effective resource for coordinating work across organisational boundaries.

Figure 7.16: review internal and external relationship networks

Checklist of key points and actions to review internal and external relationship networks

- Complete internal and external partner relationship maps (see chapter 2) and identify opportunities for improvement in relationships, communication and service delivery at an individual, team and organisational level.

- Put in place succession plans for key players and influencers to ensure that critical relationships and their networks thrive beyond the life of key people.

- Develop a high level of open, transparent communication and information sharing across seamless relationship interfaces.

- Develop internal relationship facilitators and relationship manager networks and communities of practice to support leaders, work teams and specific business improvement initiatives.

- Ensure buy-in, support and commitment are gained from all key influencers and major stakeholders.

- Identify the economic buyers, user-buyers, technical buyers and coaches for the organisation's products and services, and the strength of their network and support for your organisation or business.

- Ensure saboteurs, followers, early adaptors and innovators of change (see chapter 5) are identified.

- Develop and implement Let's Go improvement plans (see chapter 5) for high impact relationship management and change management initiatives.

- Evaluate the potential for social media technology and other IT systems, such as Facebook, Wikis, customer relationship management (CRM) systems, online relationship health diagnostics (see chapter 2), to support internal and external networks and communities of practice.

- Set up networks and communities of practice that focus on issues that are strategically important to the organisation; have established goals and deliverables; and have a support governance structure and clear executive oversight.

Develop, implement and review relationship strategies and action plans

Objective: Develop, implement and review the relationship roadmap that will manage the journey between the relationship's current state and desired future state in direct support of business goals and objectives.

Depending on the organisations' relationship type, the relationship business plan, journey management plan or action plan becomes the roadmap by which the relationship is led and managed (see figure 7.17). The strategy or plan is an important vehicle for communication with participants and key stakeholders. It brings together all the cross-functional activities and helps remove roadblocks and barriers, thereby enabling buy-in and ownership. The strategy or plan is a central document that the relationship managers can use to coordinate activities and manage relationship progress.

Figure 7.17: develop, implement and review relationship strategies and action plans

Checklist of key points and actions to develop, implement and review relationship strategies and action plans

- Consolidate the strategy and activities of the other 11 steps and align or integrate activities associated with other 0 to 10RM review mechanisms and tools, such as relationship charters (see chapter 4), Let's Go actions (see chapter 5), relationship review workshops, and business review and development meetings (BRADs).

- Obtain buy-in for the strategy or action plan from all involved parties and key stakeholders.

- Regularly review performance against the objectives, initiatives, metrics (KPIs) and agreed targets of the strategy or action plan, and communicate this to all involved parties and key stakeholders.

- Use the strategy or action plan as the basis for further innovation, and informing, engaging and empowering key people in the relationship.

- Circulate and share the strategy or plan, then seek input from all interested parties. This action will signify maturity and trust among the relationship parties.

- As a guide some basic headings for the plan could include:
  - Executive summary
  - Business environment and corporate strategy
  - Relationship scope, products and services, commercial framework terms and conditions
  - Relationship charter (see chapter 4), including vision or purpose, key objectives and guiding principles
  - Actions and metrics (KPIs) associated with key objectives and relationship performance (see chapter 4), financial and non-financial (that is, what, who owns, by when)
  - Key roles and responsibilities
  - Governance and reporting structure
  - Resource requirements
  - Opportunities for growth and improvement
  - Specific actions list with owners and due dates
  - Risks or barriers to implementation.

The six outcomes

Business must run at a profit ... else it will die. But when anyone tries to run a business solely for profit ... then also the business must die, for it no longer has any reason for existence.

Henry Ford, industrialist[10]

In other words 'earning profits (in business) is like breathing in human beings: essential for survival but not the purpose of life'.[11] Leaders of high performance relationships will continually bear this in mind when considering financial and non-financial outcomes, and building living and lasting legacies from their relationships.

The six outcomes are the sweet smell of success and justify the motivators engaged and the steps taken. They are the end result or major milestone points of the improvement process and the relationship management journey. The six outcomes are the same six key results areas (KRAs) represented on the vertical axis performance scale on the 0 to 10RM Matrix (see figure 1.4 on page 10).

Each of these six outcomes will be weighted differently for each relationship type. Typically, as an organisation moves from vendor to supplier to partner segment relationships, there will be greater degrees of non-financial KRA involvement. The six outcomes and associated KPIs may also present differently for each of the relationship parties.

Financial success

Financial success is the lifeblood of any business — the underpinning ability of a firm to pay its way, grow, invest, profit, attract investors and pay them dividends. But, as stated above, it is not the sole purpose for a company's existence.

KPI measures of financial success include profitability, return on investment, total cost improvements, revenues, volumes, outputs, overall value for money, unit costs, transaction costs, working capital, other financial indicators and associated financial ratios.

Customer and stakeholder satisfaction

Customers and stakeholders are the recipients of the products and services or have investment in their success. They are the reason suppliers exist (Principle 3) and are key to what drives any organisation and its employees.

KPI measures of customer and stakeholder satisfaction include quality, cost, schedule and service levels, response times, business case or project completion, survey results, flexibility and responsiveness.

Sustainable competitive advantage

Sustainable competitive advantage generates sustainable value for the customer, beyond the cost of creating a product or service; greater value than the price the customer is prepared to pay for it; and superior quality compared with the offerings of the competition.[12]

KPI measures of sustainable competitive advantage include market share or growth, customer or supplier loyalty or retention, referred business, percentage of available business, percentage of negotiated business, superior image and reputation, employability and career development.

Best practice implementation

Best practice includes those skills, activities, techniques, systems, processes and methods that are both efficient and effective compared with others. Best practices are the platforms upon which operational excellence is achieved; competitive advantage is sustained; and stellar customer service is delivered. Through sharing and benchmarking best practices, come next practices.

KPI measures of best practice implementation include benchmarked efficiency, reliability, availability, capacity utilisation, implementation of systems, processes procedures and the degree of operational excellence.

Innovation

Innovation is something new and improved that is useful, be it an incremental or step change breakthrough improvement. Innovation harnesses the creative spirit to turn an idea, a concept or a discovery into practical reality. New and helpful technologies, products and services result from good innovation. First mover advantage, delighted customers and financial success are the ultimate business outcomes.

KPI measures of innovation include time to market, development cycle times, number and success rate of inventions and innovative ideas, innovation investment ratios, continuous and breakthrough improvement, new paradigms engaged, new practices, tools or techniques employed.

Attitude

Attitude is a state of mind that presents as an individual's views and beliefs and in turn their behaviours and actions. At an organisational level, attitude is manifested in corporate culture and values. Either way, people are the carriers of attitude, which in turn directly affects the success or otherwise of the business.

KPI measures of attitude include behaviours displayed, trust levels developed, leadership styles, nature of communications, work practices, employee engagement, relationship health survey results, and the number and nature of good news stories that benefit the organisation.

An example of a 12/12/6 roadmap

Figure 7.18 shows a simple worked example of how the 12/12/6 roadmap can be applied practically. In this case the motivators, steps and outcomes are initiated via a Let's Go purpose statement; namely, Let's Go 'engage staff willingly'. The primary motivators that apply to this initiative are highlighted according to their impact.

Figure 7.18: an example of a 12/12/6 roadmap

Let's Go: engage staff willingly		
Motivators	**Steps**	**Outcomes (KPIs)**
All of the following: Develop trust[1] Remove hidden agendas[2] Improve communication[3] Add value Reduce total costs Resolve conflicts Provide leadership Empower people Gain commitment Development ownership Break down barriers Remove fear	Deliver engagement workshop to all stakeholder groups by agreed timeframes. Review or agree on workshop agenda(s), relationship process, progress and performance. Conduct site visits to share ideas, best practices and benchmark, and prepare for workshops and engage workshop participants. Review internal business unit relationships (quality and performance, using RAD or eRAD). Review and implement skills requirements—identify skills gaps, clarify roles and responsibilities. Develop Let's Go change management plan for sustainable long-term employee engagement.	**Attitude** Employee engagement survey, relationship health check (using RAD and eRAD results). Number and quality of engagement workshops as measured by workshop participant survey score. Number and spread of good news stories. **Innovation** Number of new ideas implemented and the financial and non-financial benefits gained. **Best practices** Number of best practices shared. Number of site visits completed plus feedback and value gained.

Note: The superscript numbers 1, 2, 3, indicate motivator priority

In this case the building of trust is seen as the most important motivator. The five steps identified in column two provide the mechanisms and structure by which the identified three primary motivators can be realised. The third column looks at the outcome areas that will be affected and identifies specific key performance indicators or measures of success.

Conclusion

So, where to from here? Simple, really: make a difference by building better relationships, in every aspect of your life.

Moving from the current state to a desired future state involves envisioning the right relationship approach to the right level of performance, and then taking the right journey to fulfil the end in mind. One thing is certain: relationships are not going away any time soon. They are here to stay and are a fundamental part of business strategy, as well as impacting the tactical, operational and day-to-day activities of everything we do.

History is written on the dreams and ideas of people who have changed the world — be that change in science, technology, business, politics, sports, arts, entertainment and the environment. In every case those people have required supporting relationships to achieve their goals. While history is a sound guide and a benchmark, the present is our opportunity to seize the day, to dream and lead desired future states that will ultimately build a positive and enduring legacy for next generations.

To that end, legacy building for a sustainable future requires broader vision, greater levels of trust and transparency, around aligned and common goals for wider and more collective benefits. Society, business and the environment in which we all live are inextricably linked. To see the connection we only have to look at the big hairy audacious goals (BHAGs) — the fundamental challenges that face current and future generations at global, organisational and personal levels. These BHAGs and fundamental challenges are shown in table 7.1.

Table 7.1: life's BHAGs and fundamental challenges for current and future generations

Global	Organisational	Personal
Global warming solutions	Visionary leadership	Leaving a legacy
Global health	Financial success	Gaining wisdom and making a difference
World peace	Integrity and credibility	Friendship and community
Solving world hunger	Right people	Hope and opportunity
Universal education	Sustainability	Safety, health and education

Each fundamental challenge, in every aspect of our lives, involves relationships, big and small, good, bad or indifferent. Relationships are core to human existence, giving us purpose and reason. They are the lifeblood by which our society functions and how we as human beings relate to each other. Tolerance, understanding, trust, attitudes and behaviours will all vary depending on the quality and performance of our relationships.

Better relationships—high performance relationships—will indeed make our world a better place. This will involve relationship rescue, relationship improvement and relationship transformation at a global, organisational and personal level. While the focus of this book has been on business relationships, the principles, models and tools can apply to every aspect of our lives. To that end, it is my hope, that 0 to 10 Relationship Management can make a fundamental and positive difference.

Remember the way:

Keep the faith

Stay focused

and

Enjoy the journey

Summary

- The 12/12/6 roadmap process outlines the details of managing relationships. The process is made up of 12 motivators, 12 steps and six outcomes

- Principle 6, Relationship management is a process, not an event, and a journey, not a destination, has direct relevance to journey management and the 12/12/6 roadmap process.

- The 12 motivators determine the 12 steps, and implementing the steps will deliver the six outcomes.

- In beginning the 12/12/6 roadmap process, you can start anywhere you like, and engage as many steps as appropriate at any time. The process is continuous.

- You can develop a relationship strategy, action or business plan using the 12/12/6 roadmap process and use it continuously to review and monitor progress.

- The five 0 to10RM themes and six principles are linked. Use the 0 to 10RM storyboard as a method of aligning activities within the 12/12/6 roadmap process and the other 0 to 10RM principles and themes.

APPENDIX A:
MASTER NEGOTIATION TERMS SHEET

This appendix provides a sample of a master negotiation terms sheet for general agreements and commercial frameworks associated with long-term or collaborative strategic relationships (that is, Type 7 Key, Type 8 Partnering and Alliancing, Type 9 Pioneering and Type 10 Community relationships). This example (see table A1) is intended to be read as a sample table and its content is intended as illustrative only. Advice from qualified legal practitioners or guidance from relationship management consultants with specialist expertise should be sought when considering any legally binding contractual or agreement framework terms. There is more information about negotiating agreements in chapter 4.

Table A1: head agreement, master agreement and general components

Major heading in the agreement	Minor heading or comments in the agreement	Terms required or to be discussed or negotiated
Form of the agreement	• Structure of the agreement	• Agree on the negotiation headings and priorities upon which the agreement and relationship will be based • Is there a master agreement and sub-agreements that cover, for example, additional work scopes, business cases and projects?
Guiding principles	• How we will work together • Communication • Co-operation or collaboration	• Establish the intent of the parties • Draft and agree on a list of principles

(continued)

Table A1 *(cont'd)* **: head agreement, master agreement and general components**

Major heading in the agreement	Minor heading or comments in the agreement	Terms required or to be discussed or negotiated
Vision and strategic objectives for the relationship	• Vision • Strategic direction, objectives and value propositions	• What is the vision and broad strategic objectives for the relationship? • What are the underlying value propositions for the relationship in delivering superior value and mutual benefit for both parties? • Is there alignment in the business drivers and objectives of each organisation? • How will you eliminate the win–lose, lose–win, lose–lose options and enhance the win–win options for the strategic relationship?
Governance structure	• Contact details • Relationship management • Performance reviews • Balanced KPI scorecard • Review of charging methods • New charges and services • Resolution of disagreements • Policy development • Leadership • Strategy development	• How will the relationship be managed—for example, roles and accountabilities of the leadership and management teams and sub-teams/third parties? • When will meetings occur? • What form of measurement and KPI scorecard will be used? • What process will be used for managing relationship issues (issue resolution)? • What roles and responsibilities and accountabilities (leadership or operational) will be for teams and individuals? • What risk management or risk sharing is planned? • What are the communication rules and protocols?
Multi-level interfaces and reporting structures	Individual roles, responsibilities, accountabilities and reporting structures	• What review of existing organisation and team structures interfaces will be conducted? • Have the roles and supporting organisational charts that support the alliance been defined and prepared?
Scope	What is in scope, for instance, for maintenance planning, turnarounds, projects?	• Have you set clear scope, with flexibility to change as needs change? • Have you organised early engagement in the planning, scheduling and design phases as appropriate and as agreed?
Exclusivity	Does exclusivity apply to all scope areas?	• What are the conditions for contracting outside parties and what is the subsequent impact on in-scope activities?

Exclusions	What is out of scope?	What is the impact on in-scope activities?
Term	<ul><li>Terms of master and sub-agreements or ancillary agreements (as appropriate)</li><li>Termination for breach</li><li>Termination without cause</li><li>Termination for convenience</li><li>Cancellation</li><li>Charges arising from cancellation</li><li>Exit mechanisms</li><li>Ongoing responsibilities beyond cancellation, such as defects liabilities</li><li>Rules of disengagement</li></ul>	<ul><li>For example, a five-year term with annual rollover for the term (that is, five years) based on agreed performance levels achieved or exceeded (that is, evergreen term based on performance)</li><li>Otherwise, a fixed or variable term, extension options etc</li><li>The rights of each party for termination, both with and without cause</li><li>Preferred exit mechanisms</li><li>Ongoing responsibilities required after termination, such as assistance to transfer to new provider, confidentiality, defects liabilities, business continuity</li></ul>
Commercial framework Pricing and remuneration	<ul><li>Performance-based model</li><li>Incentives and KPIs</li><li>Form of invoicing</li><li>Timing</li><li>Payment</li></ul>	<ul><li>What remuneration model will be used to match risk and reward to both parties?</li><li>What are the operating principles or rules of engagement surrounding the pricing and remuneration model?</li><li>How will this be measured?</li><li>Where performance is milestone based (say, technology swap out, project or shutdown delivery) what timing is associated with the milestones?</li></ul>
Principles for risk management	<ul><li>Organisation A's rights and obligations, responsibilities and accountabilities</li><li>Organisation B's rights and obligations, responsibilities and accountabilities</li></ul>	<ul><li>What are the rights and obligations of both parties? Does this go beyond Organisation A and Organisation B simply acting in good faith?</li><li>Have you identified each organisation's areas of risk management, responsibilities and accountabilities?</li></ul>
Business planning	<ul><li>Forecasting, planning, scheduling, process</li><li>Forecasting, planning, scheduling periods</li><li>Involvement of organisations in the strategy, forecasting, planning, scheduling process</li><li>Resource management</li><li>Joint business plan</li></ul>	<ul><li>What commitment will Organisations A and B make to provide open, accurate, timely access to information or to the decision making process across all scope activities?</li><li>What access will there be to good, accurate baseline data and what ongoing forecast and planning process is required?</li></ul>

(continued)

Table A1 (cont'd) : head agreement, master agreement and general components

Major heading in the agreement	Minor heading or comments in the agreement	Terms required or to be discussed or negotiated
		• What involvement and at what stage will the supplier have in this process? That is, early engagement, joint planning, scheduling, project management etc • What alignment and integration of budgets, the budgeting process, reporting systems and KPIs has been allowed for? • Has a joint business plan involving three one-year horizons been developed and what will be the basis of performance review and contract extension?
Sub-agreements, business cases, projects	• Approval process • Scope • Term • Commercial framework • Performance review process	To be reviewed on a case-by-case basis in line with the master agreement and linked to the strategic plan or forecasting process
Cultural alignment and change management	Training and development	• Induction training • Ongoing alliance training, and relationship management and change management training and development • Development of a change champions team, relationship facilitators
Reporting	Obligations for reporting for both organisations	• What are the broad obligations for reporting on both contract compliance and operational performance for both organisations? • What format, content and timing is required and to whom will reports be made on performance?
Scope change	Form of variation	• Can the agreement be changed in scope, terms and conditions? If so how? • How are variations dealt with?
Liability and insurance	• Liability – direct and consequential damages • Reciprocal liability • Capping of liability • Self-insurance • Exclusions of liability	• What liability is acceptable to both organisations or expected by both organisations? For example, will it be limited by the nature of the risk–reward model? • Are these terms reciprocal? • What capping of liability needs to occur? • What liability is excluded, beyond force majeure?

Intellectual property	<ul><li>Ownership</li><li>Existing</li><li>New works</li><li>Software Licences etc</li><li>Liability and indemnities arising from breach of intellectual property</li></ul>	<ul><li>Who owns existing and future intellectual property?</li><li>What happens in of the event of breaches of intellectual property?</li><li>How will transfer of software licences and hardware maintenance agreements be managed?</li></ul>
Issue resolution and escalation	The process	<ul><li>All issues resolved within the relationship with rights to access expert third parties, mediation, arbitration</li><li>No blame, high accountability principle at work</li></ul>
Information sharing	Confidentiality	Standard clauses
Managing other relationships	Managing third parties	<ul><li>What relationship types and performance levels are expected with third-party organisations, such as sub-alliance agreements, back to back, or other types of contractual/relationship engagement?</li><li>How will the value gained from third parties be managed?</li><li>Are management fees applicable?</li></ul>
General	<ul><li>Health and safety</li><li>Harassment</li><li>Solicitation of employees</li><li>Assignment and novation</li><li>Subcontracting services</li><li>Impact of commercial legislation</li><li>Governing law</li><li>Delivery of notices and invoices</li><li>Insurance and indemnities</li><li>Interpretation</li></ul>	Standard clauses
Managing the transition phase	<ul><li>Having the best people for the job</li><li>Transfer of people, people exchange, secondments</li><li>Skills and competency mix</li><li>Succession planning for key influencers</li><li>Taking the best for client approach</li><li>Scope and protocol for communication and information sharing</li></ul>	<ul><li>What is required to ensure uninterrupted performance?</li></ul>

(continued)

Table A1 *(cont'd)* : head agreement, master agreement and general components

Major heading in the agreement	Minor heading or comments in the agreement	Terms required or to be discussed or negotiated
Relationship performance review	Timing and process for review	• Who will review the relationship and individual partner performance? • How often will performance reviews be held? • What are the implications for better than and worse than expected performance?
Innovation	Incentives for innovation and improvement	• How will the relationship and the partners be incentivised to innovate for continuous and breakthrough improvement? • How will the benefits from innovation be monitored, measured, recognised and rewarded? • How will new ideas be captured, reviewed, implemented and the benefits measured and feedback given?

APPENDIX B: ACTIVITIES

The general competencies of high performance relationship managers

Use table B1 to score yourself between 1 and 10 (strongly disagree to strongly agree) for each criterion to assess your capabilities as a high performance relationship manager.

Table B1: checklist for rating yourself as a high performance business relationship manager

No.	Competency and performance criteria	Strongly disagree 1–2	Disagree 3–4	Neutral 5–6	Agree 7–8	Strongly agree 9–10	N/A
1	**Business effectiveness**						
1.1	I have a comprehensive understanding of the supply chain, target industries and markets, including all external influences, and have a sound understanding of competitors and their strategies.						
1.2	I have a good understanding of the internal relationships across business units and support functions and the level of their willingness and capability to meet or exceed external customer, supplier and stakeholder expectations.						
1.3	I have a good understanding of financial tools and techniques that are fit for purpose to the job role and can engage with others to develop financial strategies to protect and add value to the financial interests of my own organisation, as well as those of our customers, suppliers and stakeholders.						
1.4	I am regarded as a relationship specialist at all levels of the organisation and by external customers, suppliers and stakeholders.						
2	**Management effectiveness**						
2.1	I make a significant contribution to the development of the relationship journey management and improvement plans, and to linkages and influences in other areas of organisational strategy and general business activities.						
2.2	I can, as appropriate, take a management and leadership role over people and teams that don't report directly to me.						
2.3	I play a role that includes developing and implementing remedial plans, individually or within teams, in the event that the business relationships are at risk or other key strategic accounts need to be retrieved.						

3	**Personal effectiveness**						
3.1	I lead the relationship development process and significantly influence senior management within the organisation, and in customer or supplier organisations, in gaining their agreement to relationship strategy and improvement plans.						
3.2	I am widely consulted and respected on high performance relationship management and change management.						
3.3	I am able to turn a crisis (internal or external) into an opportunity and I am regarded as a master troubleshooter and coordinator in areas of problem solving and conflict resolution on major and critical issues.						
3.4	I have the ability to mobilise and utilise a wide range of internal and external resources for the resolution of problems, implementation of remedial plans or the development of opportunities.						
3.5	I have overall responsibility and accountability for the short-, medium-, and long-term wellbeing of nominated relationships.						
3.6	I am well versed in team dynamics and capable of managing, and effectively working within, complex and diverse team structures; however, I can act independently as required.						
4	**Customer relationships and communication**						
4.1	I can manage critical customer and supplier relationships involving all 0 to 10RM relationship types and associated performance levels with particular focus on supplier and partner segment relationships.						
4.2	I am able to extend influence and skills to other relationships as required, acting in a third-party consulting or coaching capacity.						

(continued)

Table B1 *(cont'd)*: checklist for rating yourself as a high performance business relationship manager

No.	Competency and performance criteria	Strongly disagree 1-2	Disagree 3-4	Neutral 5-6	Agree 7-8	Strongly agree 9-10	N/A
5	**New business**						
5.1	I can identify long-term strategic opportunities at a national or international level, or both.						
6	**Presentation and interpersonal skills**						
6.1	I use fully developed professional selling and broad business skills to persuade, influence, coach and facilitate at all levels from senior management to the shopfloor.						
6.2	I can exercise appropriate judgement, independently or as part of a team, on a wide range of business issues.						
6.3	I possess a clear understanding of the short-, medium-, long-term objectives and strategies, and current and future requirements for given relationships and how to achieve them.						
7	**Professional expertise**						
7.1	I am sensitive to company and customer or supplier cultures, organisational issues and market pressures, and can integrate them at a strategic level and an operational level.						
7.2	I possess high level negotiation and leadership skills and industry or market specialist skills.						
7.3	I enjoy a high level of trust, respect and credibility within allocated relationships and throughout the supply chain.						

Rating

If your score from table B1 is, on average, higher than 7 out of 10 per question, you are in the range of a high performance relationship manager.

Culture survey

Use table B2 and figures B1 to 10 to complete the survey. On a scale of 1 to 10, rate your organisation or work group (Party A) and another organisation or work group with which you have an important relationship (Party B). A score of one (1) is most like the picture on the left, while a score of ten (10) is most like the picture on the right. Ratings are discussed beneath the table. More information on the culture survey can be found in chapter 5.

Table B2: culture survey of your organisation

Culture rating	Figure	Party A	Party B
Initiative and ownership—the degree to which individuals exercise these qualities	B1		
Leadership and management—the different styles that are demonstrated	B2		
Performance recognition and reward—the behaviours and results that are valued	B3		
Managing change—by demonstrating willingness and capability	B4		
Managing conflict—by the application of process and attitude	B5		
Outward focus—the development and implementation of external strategy and tactics	B6		
Internal cooperation and collaboration—the extent to which teams are enabled and relationships are built	B7		
Employee engagement and loyalty—how employees relate to the organisation	B8		
Communication—the quality and extent of information sharing	B9		
Legacy focus—the way the organisation relates to broader, longer term social, business and environmental issues	B10		
Total (out of 100)			

From table B2 results, use the strengths and weaknesses and the areas of alignment and misalignment identified to build better understanding between the relationship parties and identify opportunities for improvement.

Figure B1: initiative and ownership—the degree to which individuals exercise these qualities

1 - 2 - 3 - 4 - 5 - 6 - 7 - 8 - 9 - 10

Figure B2: leadership and management—the different styles that are demonstrated

1 - 2 - 3 - 4 - 5 - 6 - 7 - 8 - 9 - 10

Figure B3: performance recognition and reward — the behaviours and results that are valued

1 - 2 - 3 - 4 - 5 - 6 - 7 - 8 - 9 - 10

Figure B4: managing change — by demonstrating willingness and capability

1 - 2 - 3 - 4 - 5 - 6 - 7 - 8 - 9 - 10

Figure B5: managing conflict—by the application of process and attitude

1 - 2 - 3 - 4 - 5 - 6 - 7 - 8 - 9 - 10

Figure B6: outward focus—the development and implementation of external strategy and tactics

1 - 2 - 3 - 4 - 5 - 6 - 7 - 8 - 9 - 10

Figure B7: internal cooperation and collaboration — the extent to which teams are enabled and relationships are built

1 - 2 - 3 - 4 - 5 - 6 - 7 - 8 - 9 - 10

Figure B8: employee engagement and loyalty — how employees relate to the organisation

1 - 2 - 3 - 4 - 5 - 6 - 7 - 8 - 9 - 10

Figure B9: communication—the quality and extent of information sharing

1 - 2 - 3 - 4 - 5 - 6 - 7 - 8 - 9 - 10

Figure B10: legacy focus—the way the organisation relates to broader, longer term social, business and environmental issues

1 - 2 - 3 - 4 - 5 - 6 - 7 - 8 - 9 - 10

Rating

- If your total score is higher than 80 out of 100 and you scored more than 7 out of 10 in every category, your organisation is in good shape for high performance relationship management in the partner segment.

- If your organisation's score is between 50 out of 100 and 80 out of 100 and you scored more than 5 out of 10 in every category, your organisation is in the supplier segment at varying performance levels or in high performing vendor relationships.

- If your organisation's score is less than 50 out of 100, it is more comfortable with vendor relationship approaches.

APPENDIX C: TEMPLATES

This appendix provides a draft relationship review and improvement workshop agenda that can be modified to meet specific desired workshop outcomes.

Template for a relationship review and improvement workshop

Participating companies, departments or functions:

Date:

Location:

Workshop purpose

To review the current state and desired future state for the relationship and jointly agree how to bridge the gap in direct support of agreed business goals and objectives.

Desired outcomes

- To understand or reconfirm the business alignment between the two organisations, departments or functions (that is, business drivers, priorities and objectives).

- To review the highlights, lowlights and rub points for the relationship to date, and identify the opportunities for improvement and barriers to implementation moving forward.

- To review high performance, collaborative relationship management (principles and practice).

- To conduct a relationship health check to gain understanding of and agreement on the current state and future desired state for the relationship.

- To develop a relationship charter (that is, shared vision, joint key objectives and guiding principles) and the associated measures of success or key performance indicators (KPIs).

- To agree on and document an action plan to manage future opportunities for improvement and deal with any barriers to implementation.

- To review governance, communications, systems and processes that support the relationship.

- To walk away from the workshop with a common understanding, joint commitment, and a clear roadmap for relationship development and improvement.

Attendees and roles

Chairperson, sponsor, convener or leader	
Facilitator(s)	
Workshop meeting participants	
Pre-work (e.g. reading, preparation, surveys, interviews, meetings, audio visual requirements, book venue)	

Agenda for the day

Time	Agenda item	Objective or process	Outcomes
8.30 – 8.45	1. Introduction and opening remarks *Who:* Facilitator	To set the scene: • Background to the day. • Overview of process to be taken and desired outcomes. • Introduction of participants to each other. • Agree on ground rules for participation.	
8.45 – 9.15	2. Review alignment of vision, strategies, business drivers for the relationship parties *Who:* Facilitator	• To present an overview of the current business background and climate, strategic objectives, business drivers, issues and opportunities from senior executives' perspective.	• A better understanding of the business priorities and the degree of alignment between the relationship parties. • An appreciation of the need for the workshop and the role the participants can play.

8.45 – 9.15		• To understand why this relationship is important to the success of all the parties from the perspective of strategic value and commercial value. • To agree on the value propositions underpinning the relationship. • To share information—Q&A response from participants.	
9.15 – 10.15	3. Review of the relationship practices and performance to date *Who*: Facilitator	Group or split group exercise to: • Understand some of the history and current state of the relationship: - highlights - lowlights - rub points. • Review: - opportunities for improvement - barriers to implementation, covering areas including commercial, service and performance levels; relationship and people aspects and interfaces; and best practice and innovation.	• An agreed perspective of the past and the present operating environment, and future opportunities. • To be in a position to move forward in an innovative and collaborative manner.
10.15 – 10.30	Break		
10.30 – 11.00	4. Review of high performance relationship management (0 to 10RM), principles, concepts and practices *Who*: Facilitator	• To review 0 to 10RM storyboard principles and themes. • To achieve a common understanding of the principles, concepts and practices involved in high performance relationships, such as: - shared vision, common goals - shared risks and benefits, trust - performance-based outcomes - openness and transparency - joint teamwork. • To understand the role of relationship champions, key influencers and doers, and sponsors. • To agree on what trust looks like for the relationship.	• Joint agreement to proceed to the next steps. • Common understanding agreed. • Areas of conflict and alignment agreed. • Additional opportunities for improvement documented.

(continued)

Time	Agenda item	Objective or process	Outcomes
11.00 – 12.00	5. The relationship health check using the relationship alignment diagnostic (RAD), an intuitive analysis, or the eRAD *Who*: Facilitator	• To understand the degree of alignment between the relationship parties on culture, strategy, structure, process and people. • To review RAD or eRAD results based on the current state of the relationship and the desired future state as identified on the 0 to10RM Matrix (i.e. what relationship type and performance levels are desired and how to get there). • To identify future value-add opportunities and key objectives for the relationship.	• RAD completed, that is, current state and future desired state identified and discussed. • Opportunities for improvement and key objectives documented.
12.00 – 12.45	Lunch		
12.45 – 2.00	6. Develop or review the relationship charter *Who*: Facilitator	• To understand the principles and intent behind the relationship charter or review the existing charter. • To develop or review the relationship charter, including: - vision statement or mission statement - key results areas (KRAs) and key objectives (8–12) - guiding principles. • To understand the links to finance and organisational structures, and key performance indicators (KPIs).	• A signed relationship charter and the joint commitment and ownership for successfully implementation.
2.00 – 3.00	7. Review of KPIs for the relationship 8. Develop horizon-based action plan(s), such as 100 days, medium term and long term *Who*: Facilitator	• To review existing KPIs if appropriate. • To agree on future KPIs (or draft) for the relationship (i.e. from the key objectives and opportunity areas identified). • To understand the link between performance, measurement and remuneration for the relationship and attitudes of people.	• Documented existing KPIs if appropriate. • Agreed on set of draft KPIs. • Action plan(s) developed.

2.00 – 3.00		• Develop Let's Go action plans or strategies that surround the joint key objectives and associated KPIs, on 100-day horizon, and medium to longer term actions and strategies: - Where (from and to) - Why - What - Who - When (start and finish)	
3.00 – 3.15	Break		
3.15 – 3.45	Agenda items 7 and 8 *continued*		
3.45 – 4.45	9. Agreeing on some details *Who*: Facilitator	For example: • To review the makeup and mechanics of current commercial arrangements. • To review relationship governance. • To review the role of the relationship manager(s) and operational teams. • To identify individual/ team roles and concerns. • To review the issue resolution and escalation process. • To review communication strategy/ protocols. • To agree on and draft a workshop outcomes summary list from the day (i.e. an elevator speech).	• Agreement on all issues and document as appropriate.
4.45 – 5.00	10. Sign off on the next steps and close the meeting *Who*: Facilitator	• To confirm participants support and commitment for the next steps and their roles from this point on. • To review the value gained from the day.	• Agreement and documented actions on who, how, when and why to proceed. • Support for and commitment to the next steps. • One team, one direction, common goals agreed.

Sample workshop and meeting norms or ground rules

Add to or delete from this list as is agreed or appropriate.

- Arrive at or before start time.

- Start every meeting with a safety moment (a safety story to share learnings).

- Introduce meeting participants.

- Engage in open, honest, relevant, timely discussion.

- One voice at a time.

- Challenge with respect.

- It is okay to disagree, but it is not okay to be disagreeable.

- All problems have solutions.

- Be proactive in seeking win–win solutions.

- Park and process the details.

- If you are assigned actions or responsibilities, deliver on your commitments IFOTA1 specification.

- Phone off or at least on vibrate mode.

Template for a relationship health check

You can use this blank relationship diagnostic (RAD) chart (see figure C1) to conduct a relationship health check on your relationship with another organisation.

Figure C1: the relationship diagnostic (RAD)

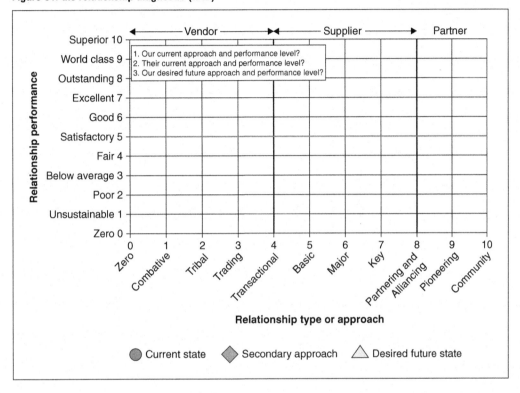

REFERENCES

Chapter 2: The framework

1   M Gandhi, *The golden treasury of wisdom — thoughts and glimpses of life*, Anand Limye India Printing works, Mumbai, 1995, p. 37.
2   SR Covey, *The seven habits of highly effective people*, The Business Library, New York, 1990, p. 95.
3   T Lendrum, *The strategic partnering handbook*, 4th edn, McGraw-Hill, Sydney, 2003, p. 164.
4   G Hamel & CK Prahalad, *Competing for the future*, Harvard Business School Press, Boston , 1994, p. 223.

Chapter 3: The 11 relationship types

1   T Levitt, *The regis touch*, Addison-Wesley, Menlo Park, CA,1986, p. 35.
2   T Peters, *Thriving on chaos*, Pan Books, London, 1989, p. 278.
3   T Lendrum, *The strategic partnering handbook*, 4th edn, McGraw-Hill, Sydney, 2003, p. 287.
4   R McNeish, 'Lessons from the geese'. The lessons were part of a sermon delivered by poet and teacher Robert McNeish in Baltimore in 1972.
5   Grameen Bank 15 May 2011, viewed 29 May 2011, www.grameen-info.org/index.php?option=com_content&task=view&id=632&Itemid=664.
6   Interview between reporter Peter Day and Alberto Vollmer, BBC News Podcast report 18 August 2009.
7   T Lendrum, *The strategic partnering handbook*, 4th edn, McGraw-Hill, Sydney, 2003, p. 426.

8   Fonterra, viewed 29 May 2011, www.fonterra.com/wps/wcm/connect/
    fonterracom/fonterra.com/our+business/fonterra+at+a+glance/about+us/key+facts.
9   Habitat for Humanity, viewed 29 May 2011, www.habitat.org/how/historytext.
    aspx and www.habitat.org/how/factsheet.aspx.

Chapter 4: Key components

1   T Lendrum, *The strategic partnering handbook*, 4th edn, McGraw-Hill, Sydney,
    2003, p. 56.
2   D Goleman, 'Leadership that gets results', *Harvard Business Review*, vol. 78, no. 2,
    2000, p. 78.
3   T Lendrum, *The strategic partnering handbook*, McGraw-Hill, Sydney, 2004,
    pp. 214–18.
4   R Whiteley, *The customer driven company — moving from talk to action*,
    Business Books, London, 1991, p. 26.
5   RS Kaplan & DP Norton, *The balanced scorecard*, Harvard Business School Press,
    Boston, 1996, p. 24.

Chapter 5: Let's Go change model

1   R Kipling, 'I keep six honest serving men', in 'The elephant's child', *Just so stories*
    (originally published Macmillan & Co., 1902), Wikipedia, viewed 29 May 2011,
    http://en.wikipedia.org/wiki/Just_So_Stories.
2   Thinkexist.com, viewed 29 May 2011, http://thinkexist.com/quotes/
    charles_darwin/.
3   R McKenna, *Relationship marketing*, Addison-Wesley, Don Mills, Ontario, Canada,
    1991, p. 114.
4   J Barker, 'Paradigm pioneers', *Discovering the future series* (video), ChartHouse
    International Learning Corporation, Burnsville, MN 1993.
5   JP Kotter & James L Heskett, *Corporate culture and performance*, The Free Press,
    New York, 1992, p. 47.
6   TG Harris, 'The post-capitalist executive: an interview with Peter Drucker',
    Harvard Business Review, 1993, pp. 114–24.

Chapter 6: The relationship development curve

1   RP Lynch, *Business alliances guide*, John Wiley, New York, 1993, p. 279.
2   A Noble, viewed 25 May 2011, http://en.www.thinkexist.com/authors/
    Alex_Noble.

3   J Barker, 'Paradigm pioneers', *Discovering the future series* (video), ChartHouse International Learning Corporation, Burnsville, MN, 1993.

4   GB Shaw, *Back to Methuselah*, Project Gutenberg eBook, 2004 (originally published 1921), p. 41

5   T Levitt, *The regis touch*, Addison-Wesley, Menlo Park, CA, 1986, p. 35.

6   JD Lewis, *The connected corporation*, Free Press, New York, 1995, p. 106.

Chapter 7: The 12/12/6 roadmap process

1   Lao-Tzu, *The book of the way*, Kyle Cathie Limited, London, 2000, p. 64.

2   C Powell, *A soldier's way — an autobiography*, Hutchinson, London, 1996, p. 264.

3   M Walton, *The Deming management method*, Mercury Books, London, 1989, p. 72.

4   J McConnell, *Safer than a known way*, Delaware Books, Sydney, 1988, p. 224.

5   S Covey, *The seven habits of highly effective people*, Free Press, New York, 1989, p. 235.

6   J Katenback & D Smith, 'The discipline of teams', *Harvard Business Review,* 1993, p. 110.

7   P Scholtes, *The team handbook*, Joiner Associates, Madison, 1991, pp. 6–11.

8   J Barker, 'Paradigm pioneers', *Discovering the future series* (video), ChartHouse International Learning Corporation, Burnsville, MN 1993.

9   G Hamel & CK Prahalad, *Competing for the future*, Harvard Business School Press, Boston , 1994, p. 321.

10 FF Reichheld, *The loyalty effect*, Harvard Business School Press, Boston, 1996, p. 16.

11 N Stone, 'Editorial', *Harvard Business Review*, vol. 75, 1997, p. 14.

12 ME Porter, *Competitive advantage: creating and sustaining superior performance*, The Free Press, London, 1985, p. 3.

INDEX

Printed in Australia
02 Nov 2018
688790